THE THEISTCIDEIST

A TRANSHUMANIST EXPLORES RELIGION, SPIRITUALITY, AND ATHEISM

ZOLTAN ISTVAN

AUTHOR'S NOTE

While these essays have been arranged and edited for readability, many of them appear similar (if they are not new) to how they were originally published. Attempts have been made to preserve the context and moment in time they were written. Some articles contain British spelling. Publishing information and fact checks can be found by utilizing the Appendix.

TABLE OF CONTENTS

INTRODUCTION

CHAPTERS

VI: Secular Visits & Events

****

INTRODUCTION

For years I've called myself an atheist, but naturally I am not one. Anyone that thinks they know for sure whether God exists or doesn't exist is taking an irrational stance in not thinking about the question probabilistically. Some things are simply unknowable by the three pounds of meat called a brain we all carry on our shoulders—there's zero evidence today one way or the other to prove divinity (or lack of it).

That said, while agnosticism is the most rational perspective to hold, there's ample evidence that institutionalized religion through the ages has harmed humanity greatly. The Inquisitions, multitude of holy wars, September 11th attacks, child brides, cliterdectomy—and even minutiae like my own personal suffering at Catholic grade school—are just a few of the countless pieces of evidence that show religion harms society. It's for this reason I sometimes actively take an atheistic stance in public life, and not the more docile agnostic path. Atheism for me is a platform to challenge the irrational and harmful beliefs the majority of the world insists upon, whether it be blue four-armed deities, Arabic prophets, crucified saviors claiming to be the son of God, or the thousands of other divinities humans have contrived.

For most of my adult life, I've often considered myself an apatheist, one who doesn't care about religion or gods in one way or the other. But modern physics and cosmology has made me leave that badge behind. Too many marvels of science and technology make me realize that we are almost certainly not alone in the universe—and that super-intelligent alien civilizations probably comprised of artificial intelligence or even pure organized atomic intelligence are scattered through the cosmos. I'm not alone in this once-kooky belief; plenty of NASA scientists thinks the same thing now. The evidence is simply overwhelming.

The university has trillions of stars and billions of habitable planets. In all likelihood, we are not the sole accident of intelligent life in the nearly endless galaxies out there. That would be a near mathematical impossibility, and it also sounds like modern-day Flat Earth theory to me.

But then what if we are not alone? Is God out there? Or is something out there quickly evolving into God other than embodied AI? Or did God already exist and die, as I explain in my controversial theistcideist perspective—which is the best theory I have right now supporting my agnostic metaphysical and personal beliefs.

It's exciting to ponder what new powerful intelligences will be like in the 21st Century, especially as humans are on the cusp of creating bonafide artificial general intelligence—something that will almost surely rival and quickly surpass our brain power in the next decade or so. What will an intelligence be like when it's twice as smart as people? What about ten thousand times smarter?

The exponential evolution of the microprocessor will likely lead to a Singularity event probably by 2050 when our three-pound brains can no longer understand the evolved intelligence of super-smart machines. They will simply leave humans behind in terms of logic, understanding, and experience. For theists and atheists alike, the million-dollar question is: *How smart does a machine need to be to be called God?*

I don't have the answers to this and other religious-futurist questions. But my book of essays below attempts to organize my years of writing and campaigning as a secular activist and religious thinker. While I'm steadfast in my dedication to reason and the Scientific Method—and to ending the harm formal religion has caused to society—the essays below also explore what the future of atheism and, yes, spirituality, may be like in the coming transhumanist age.

Zoltan Istvan / January 5, 2020

CHAPTER I: FUTURE OF ATHEISM

1) Do We Have Free Will Because God Killed Itself?

Some people believe humans with our three-pound brains are the most advanced life form ever to exist; I am not one of them. To insist we are alone in the universe, or that we are the galaxy's crowing civilization, reeks of ego—and reminds me of those who insisted the Earth was flat.

The universe is 13.8 billion years old according to experts. A lot can happen in so much time, such as the rise (and fall) of superintelligences amongst the approximately two billion life-friendly planets that exist in our galaxy.

It is likely that these highly advanced intelligences long ago reached what we call the singularity: a moment in time when technological acceleration—most likely through the creation of artificial superintelligences—becomes incredibly rapid.

This presents a thorny issue to humans because of what I call the Singularity Disparity—the idea that whoever reaches the singularity first will make sure no one else can achieve a similar amount of power.

If we are not alone in the universe and also not the most intelligent life forms, then it's unlikely our species can evolve beyond a certain point, since other more advanced life forms won't allow it.

So where does that leave us, a species about 20 to 40 years away from building superintelligences that will help us reach the singularity? The answer is not pretty. In fact, if I had to guess—based on some of the recent discoveries in string theory—we're likely already existing in some type of simulation created by an ancient superintelligence, one where we're observed, regulated, and possibly even manipulated at times.

Worse, other superintelligences likely structured the intelligences controlling us before them, and so on.

I'm not going to argue the merits of whether we live in a simulated hologram universe or not, all of which have been covered by philosophers through the ages, from Aristotle to Oxford's Nick Bostrom to John Searle and his Chinese room. Suffice to say, there's enough scientific and philosophical evidence for me to slightly tilt in favor of it all. For me, however, the more interesting question is why would we live in a simulation? Given the Singularity Disparity, why would some superintelligence or group of superintelligences do this to us?

There are various explanations. The main ones are:

1) We are experiments and playthings for those superintelligences using us to further understand themselves or support some causes of theirs, including dealing with boredom.

2) We are literally already intrinsic parts of those superintelligences and exist simply as their thoughts, energies, or structure (the Gaia people love this idea).

3) We are accidents in the universe and our existence is totally arbitrary.

The deity-averse existentialist in me likes #3 best, but I'm still not satisfied with any of the answers, mainly because none of them address what happened to the very first superintelligence, an entity who may have set all the universe's rules up.

Luckily, there is a fourth, more controversial take that I do think is worth exploring. The foundation of the universe, including all the simulations, probabilities, and possibilities of existence are the result of the first and most powerful superintelligence killing itself.

In short, an entity literally on the verge of becoming God knowingly and willingly died by suicide.

The problem with being God—a truly omnipotent being—is that of free will. As a recent comedy skit called Future Christ on The Daily Show with Jon Stewart—a skit which partially resulted from my original atheistic story—pointed out: "If God wants to quit smoking, can he hide cigarettes from himself?"

Being all-powerful is a very strange, ironic dead end. The only thing omnipotence can truly equal is a total mechanistic void. Achieving omnipotence is literally the act of suicide, in terms of forever self-eliminating one's consciousness. This is because a conscious intelligence, as reason dictates, is based on ability to discern values—values, for example, to know whether as an all-powerful being, one can create something so heavy that one can't lift it. Values require choice. But omnipotence means that all choices have already been made, and nothing can ever change, because all variables are already accounted for and no randomness or anomalies exist.

It's quite possible, a long time ago, that the first superintelligent Singularitarian decided to up its game and attempted to become omnipotent. But if it succeeded—and it may have— then it would have become an entity without a singular intelligent consciousness, because intelligence requires choice. For all practical matters, it would cease to exist in a personal interactive way that any intelligence could relate to.

But before this first Singularitarian did that, it would've left us with its rules—physical laws of the universe that contain our potential power and intelligence. It would've also left us with the code of the Singularity Disparity, where the singularity we achieve will never equal other singularities or be the most powerful.

If this is all the case, this leaves the human race in a precarious position. Here we are, in a universe where many singularities have almost certainly taken place, but reaching anything beyond a certain point becomes impossible due to limits of pre-existing natural laws. Adding to the mix are other superintelligences that don't want us to dominate or overpower them, either, just as we don't want any other entities on Earth to dominate or overpower us. Hierarchies and power plays exist everywhere—they are the fabric of the universe.

As an atheist (or even a possible theistcideist—one who believes God or a supreme being once existed but no longer does because it terminated itself), I would commend this leading superintelligence for destroying its conscious self. By doing so, and establishing that nothing else could ever become as powerful as itself, it would've forever sown choice into the universe, since no one can ever reach a perfect position of choice-less omnipotence, and the death of its consciousness would mean it couldn't ever change what it had done. This superintelligence's final acts have assured all other advanced life forms the possibility of free will and the ability to try to become more than we are.

****

2) Mind Uploading Will Replace God

I hear a lot of philosophical complaints suggesting that being alive in a computer as an uploaded version of oneself is quite different than being alive in the physical world. While that is open for debate, one aspect of the issue people often forget about is that the so-called spirit world of Abrahamic faiths—which approximately four billion people follow—is based on something at least as odd as the bits in software code that will make up any virtual existence.

When you think about it, trying to wrap your brain around how digital technology and all its wonders are even possible is simply bizarre. Only a tiny fraction of the world's population understand such things in any depth. And an even smaller amount of people actually know how to design and create the microchips, circuit boards, and software that constitutes this stuff in the real world. Human beings are a species dependent on a tech-imbued lifestyle that none of us really understand, but accept wholeheartedly as we go on endlessly texting, Facebooking, and video conferencing.

As a non-engineer atheist grappling with the implications of 1s and 0s manifesting all digital reality, I have at least this much in common with religious people—because they can't understand the spirit world either, even if they insist it exists.

The major difference between the religious spirit world and the digital world is that engineers—technology's high priests—can recreate software, microchips, and virtual environments again and again. They can also test, view, change, manipulate, and most importantly, improve upon their creations. They can apply the scientific method and be assured that the worlds they built of bits and code exist—as surely as we know the Earth is spinning, even if we can't feel it.

People of the planet's major religions can't do this with their spirit worlds. They can only make leaps of faith, and elaborately describe it to you. One either agrees or disagrees with them. Amazingly, proof is not necessary to them.

Being able to upload our entire minds into a computer is probably just 25-35 years off given Moore's Law and the current trajectory of technology growth and innovation. If we can harness the power of artificial intelligence in the next 15 years, then we might get there quicker, as AI will likely make the research and progress happen far more rapidly. But mind uploading is generally considered possible by experts. After all, humans are just material machines, striving to create other machines that mirror ourselves and desires.

Already, interaction between microchip and the brain are occurring all around the world in the form of cranial implants helping the deaf, blind, and mentally ill. Furthermore, telepathy, accomplished last year between people in India and France, is the communication medium of the future. We're just in the infancy of all this, but progress is accelerating. I'm looking forward to having an exact copy of myself online one day, both as a companion and as a form of personal immortality in case my biological self dies.

Atheists may not believe in God, but as Sam Harris' recent bestseller, *Waking Up: A Guide to Spirituality Without Religion* points out, we are still deeply spiritual creatures, searching for answers, trying to do good upon the world, and pondering the mysteries of the universe. All this is very healthy, and not that different than some core hopes of the religious-minded. In fact, the only real difference between religious people and atheists is the fact that religious people insist an all-knowing deity is outside of themselves and controlling the shape of the world. Atheists see no God and believe unconscious universal forces and human will are responsible for the shape of the world.

It's that shape of the world that I care about. It's that shape that affects our lives and gives form to our society, nations, and deeds. For millennia, society has been controlled, guided, and manipulated by religion—often for the worse. As a rule, the more fundamental a particular religion was, the better it steered its populace in the direction the leaders of the religion wanted. I often refer to this steering as baggage culture, pieces of social structure, cultural conditioning, and archaic rules passed on from generation to generation with little philosophical change or growth, despite the fact that society evolves every year. Eventually, such baggage culture weighs us down so much that society becomes lethargic and hopelessly burdened with nonsense in its many actions. This can be seen in the United State's monopolistic two-party pretend democracy system. It can also be seen in Islam—one of the world's fastest growing religions—whose main sacred text, the Koran, is often seen as being at odds with basic modern day women's rights. Of course, one of the most embarrassing examples of baggage culture I know of is America's Imperial measurement system, which favors obfuscation instead of the better metric system.

So what can we do to eliminate our baggage culture? I'm afraid that little will happen as long as we are exclusively biological. Our instincts for vice, petty behavior, and superstition are too strong. There has certainly been a shift towards moral fortitude, reason, and irreligiosity in many developed countries, but it is slow, very slow. The sad truth is we'll be uploading ourselves into machines long before rationality and agnosticism become truly dominant on Earth. The good news, though, is as people begin uploading themselves, they'll also be hacking and writing improved code for their new digital selves—resulting in "real time evolution" for individuals and the species. It's likely this influx of better code will eliminate lots of things that, historically speaking, religion has attempted to protect people from— namely stupidity and social evil.

Take Andreas Lubitz, the co-pilot who likely intentionally crashed Germanwings Flight 9525 in the Swiss Alps, tragically killing all the people aboard. Lubitz is suspected to have been suffering from depression. In the future, we may all have avatars—perfectly uploaded versions of ourselves existing in the cloud or in chip implants in our brains—and these avatars will help guide us and not allow us to do dumb or terrible things. In the Germanwings plane incident, the avatar would have been able to eliminate the depression in itself, and then could've conveyed that to the other, real life self. In this way, the better suited person would've have been given the task of flying the plane.

This may serve what Abraham Lincoln called the better angels of our nature, which we all have but often don't use. Now, with digital clones participating in our every move, someone trustworthy will always be in our head, advising us of the best path to take. Think of it in terms of a spiritual trainer—or even guru—leading us to be the best we can be.

A good metaphor or comparison of this type of digital assistance will already be happening in the next few years when driverless cars hit the road. In the same way driverless cars will help lessen drunk driving, perfected uploaded avatars will also lead us to be more judicious, moral, and reasonable in our lives.

This is why the future will be far better than it is now. In the coming digital world, we may be perfect, or very close to it. Expect a much more utopian society for whatever social structures end up existing in virtual reality and cyberspace. But also expect the real world to radically improve. Expect the drug user to have their addictions corrected or overcome. Expect the domestic abuser to have their violence and drive for power diminished. Expect the mentally depressed to become happy. And finally, expect the need for religion to disappear as a real-life god—our near perfect moral selves—symbiotically commune with us. In this way, the promising future of atheism and its power will reside in achieving this amazing transhumanist technology. Code, computers, and science will one day replace formal religion and its God, and we will be better as a species for it.

****

3) Upgrading Religion for the 21st Century: Christianity is Forcibly Evolving to Cope with Science and Progress

Recently, the pope made history when he told his flock to accept divorced Catholics. A month later, NPR reported a gay preacher had been ordained as a Baptist minister. Next year it might well be evangelicals in the deep South turning pro-choice. Everywhere around us, traditional Christian theology and its culture is breaking down in hopes of remaining relevant. The reality is with incredible scientific breakthroughs in the 21st century, ubiquitous information via the Internet, and an increasingly nonreligious youth, formal religion has to adapt to survive.

But can it do so without becoming obsolete? Perhaps more importantly, can Christianity — the world's largest religion with 2 billion believers — remain the overarching societal power it's been for millennia? The answer is not an easy one for the old faith-driven guard.

To remain a dominant force throughout the 21st century, formal religion will have to bend. It will have to adapt. It will have to evolve. Hell, it will have to be upgraded. Welcome to the growing impact of Christian relativism.

The familiar term cultural relativism was coined by anthropologist Franz Baos, who suggested that people have a difficult time understanding another's culture without having grown up in it—so therefore we should strive to empathize more with foreign cultures and people. It's a great concept, and after many years reporting for National Geographic in dozens of countries, I came to strongly believe in the idea.

Christian relativism, however, does not have that honor of generating empathy so easily—at least not until it separates itself from its cornerstone philosophy: adherence to the Bible. Even with its many dozens of translations, most everything in the Bible simply cannot be logically interpreted in a multitude of ways—or flippantly passed over in generous empathy. To make the Bible's deity-approved instructions and ideas soundly work, church leaders pushing Christian relativism may simply have to back down or say it made a mistake with its past fundamentalism.

For example, if the Bible clearly says being gay is a sin (and it does many times), then Christians can't just wake up one day and say homosexuality is permissible without dismissing God's word. Another example is women; if the Bible says they can't be priests and must submit to men, then the church can never profess to believe in equality—which is does all the time.

Additionally, if committing blasphemy (striving to become god-like) through transhumanism is an unforgiveable sin that leads to eternal punishment, then Christians can't say they represent a loving and kind God. The hypocrisy is too much to pretend one is being logical or reasonable—since transhumanists vocally aim to never die and possibly even become gods (or God) through science and technology.

This is the dilemma that the Abrahamic religious face in the age of Christian relativism. They have sealed themselves in the ideological fort for protection, and now they have no way out without atheists and agnostics chiding them. Language is fiercely mechanical— and in the case of the Bible, many of the truths are prominently black and white.

The antithesis of the Bible is, of course, the much simpler Western ideology: the scientific method, upon which the other part of modern humanity's culture was built upon—the one that brought us skyscrapers, CRISPR gene editing tech, robots, and vaccinations so our children don't die from measles. The scientific method states nothing is black and white, but if you prove something enough times, it's safe to trust it until something strange or unwanted occurs. It's humble at its core, unlike Christianity which claims to be under guidance of an omnipotent God.

Consider Christianity's core message: You are born in sin, and only through Christ can you be redeemed and reach a happy afterlife. The scientific method would've never entertained such a conclusion, because it would've been stuck asking what is sin?, and where is Christ? — neither of which can be proven one way or the other.

With this in mind, how does Christian relativism then expect to be taken seriously? I wish that was the question, but people are so entrenched into Judeo-Christian culture, that we rarely consider that Christianity is even changing. We only think we are becoming more open-minded, and that God and our religious brethren should pat us on the back for our newfound wisdom.

While I shake my head in disbelief at the Christian mirage all around me (and the billion people who call the pope wonderfully progressive despite his disdain of condoms and other birth control), I accept it as a better fate that the far more dogmatic one humans endured in the 20th Century. I believe I speak for the one billion nonreligious people out there when I say I'll take progress however I can get it — even if it results in a Jesus Singularity, where even the superintelligent robots engineers are trying to make may end up being programmed to believe in Christ. But Christian relativism is not a cure to the disease—it's just a band-aid of belief. The cure — or better put: the sobering tonic—is the scientific method, a simple philosophy that says: Get used to not knowing anything for sure — then make up your own mind on what you believe.

****

4) Religion is Harming Society and Lives

All around the world, religious terror is striking and threatening us. Whether in France, Turkey, London, or the USA, the threat is now constant. We can fight it all we want. We can send out our troops; we can chip refugees; we can try to monitor terrorist's every move. We can even improve trauma medicine to deal with extreme violence they bring us. But none of this solves the underlying issue: Abrahamic religions like Christianity and Islam are fundamentally violent philosophies with violent Gods. Sam Harris, Richard Dawkins, Christopher Hitchens and others have all reiterated essentially the same thing.

Consider these verses from the Koran:

Koran (3:56): As to those who reject faith, I will punish them with terrible agony in this world and in the Hereafter, nor will they have anyone to help.

Koran (8:12): I will cast terror into the hearts of those who disbelieve. Therefore strike off their heads and strike off every fingertip of them.

And then consider these verses from the Bible:

Deuteronomy 17:12: Anyone arrogant enough to reject the verdict of the holy man who represents God must be put to death. Such evil must be purged.

Numbers: 31:17: Now therefore kill every male among the little ones, and kill every woman that hath known man by lying with him.

Of course, both the Koran and Bible have passages that highlight kindness too—but you don't get a get-out-of-jail-free card in the 21st Century by being both violent and peaceful. If you beat your spouse, you're an abuser and can face jail time (even if you're a loving spouse other times). It's one or the other in the 21st Century: If you're a warmonger, murderer, or a terrorist—you're a bona fide warmonger, murder, or terrorist. And nothing is going to change that.

The fundamental problem with religion is that believers—about 5 billion people right now on Planet Earth—are so sure they're "correct" on anything and everything they believe. This is, of course, a sure sign of insanity—especially since most of what people believe was taught to them when they were children (and they had no way to filter it out or reason about it).

The only real truth out there, at least while our brains are just three pound bags of meat (and our senses—like our eyes—see just 1 percent of the visible universe), is to know "absolute truth" is something way too complex to understand. The only real thing to understand right now is the Scientific Method—the holy grail of wisdom that reason advocates follow. It states that if you test a hypothesis enough times, and the outcome seems to always be similar, then you can utilize that as a semi-truth and apply it functionally in one's life (but beware: It could change anytime and it might). That's the language of reason—the language of science.

It's the same method of thinking that explains why jet airplanes don't fall out of the sky. Or why skyscrapers keep standing through hurricanes. Or why we could put a man on the moon and bring him back.

However, it's not the thinking method that President Obama used to swear on a Bible to get his job on inauguration day. Or George W. Bush when he stopped life saving stem cell funding for seven years during his presidency. Or the Pope when he insists condom usage is a sin, despite it having the possibility of saving millions of lives from AIDS in Africa.

The Scientific Method is also not the thinking method of the pilots who flew into the World Trade Center. Or of the murderer who gunned down people in Orlando. And it's certainly not the method of thinking that the truck driver used to run down innocent people in Nice, France.

Like the hundreds of millions of other nonreligious people out there, it's hard for me to fathom how religious people got brainwashed into being this way—this ignorant. But bear in mind, it's not just religious terrorism that is literally killing us—it's much more.

Consider how many nonreligious secular people there are leading our nation right now. The answer is astonishing: It's zero (at least publicly). All 535 members of Congress, all eight Supreme Court justices, and our President believe in God and an afterlife.

No wonder life extension and anti-aging science is basically unfunded by the US Government. Why should the US care about whether you live longer or can overcome disease when you're all going to wake up in Jesus' arms after you die? Or in some heavenly Islamic paradise with a bunch of virgins?

I'm a political candidate that wants you to live—not in some unknown paradise once you die that no one has ever seen before or can prove exists. I want you to live now, regardless what craziness or tragedy the world can throw at you. I want your loved ones to live too—and not die because of aging, disease, or terrorism. I want you all to survive as long as you want—and to try to find a perfect world here on Earth. Transhumanist science can give that to us, and it will soon. And maybe in a hundred years, we can all even venture somewhere else in the universe when space travel can get us there safely.

If you want to live—and not be killed or die—make a point to criticize and disavow religion and religious people for being deathist: the idea that death is either welcome or acceptable (whether it comes via terrorism, disease, or aging).

In the 21st Century, fundamental religion is a form of mental disease. And sadly, that disease continues to take lives everywhere, in the worst of ways.

****

5) To Ensure a Future of Transhumanism, Atheists Should Confront the Deathist Culture Religion Has Sown

In the West, atheism is growing. Nearly a billion people around the world are essentially godless. Yet, that burgeoning population faces an important challenge in the near future—the choice whether to support far longer lifespans than humans have ever experienced before. Transhumanism technology could potentially double our lifetimes in the next 20-40 years through radical science like gene editing, bionic organs, and stem cell therapy. Eventually, life extension technology like this will probably even wipe out death and aging altogether, damaging one of the most important philosophical tenets formal religion uses to convert people: the promise of being resurrected after you die.

About 85 percent of the world's population believes in life after death, and much of that population is perfectly okay with dying because it gives them an afterlife with their perceived deity or deities—something often referred to as "deathist" culture. In fact, four billion people on Earth—mostly Muslims and Christians—see the overcoming of death through science as potentially blasphemous, a sin involving humans striving to be godlike. Some holy texts say blasphemy is unforgivable and will end in eternal punishment.

So what are atheists to do in a world where science and technology are quickly improving and will almost likely overcome human mortality in the next half century? Will there be a great civil rights debate and clash around the world? Or will the deathist culture change, adapt, or even subside? More importantly, will atheists help lead the charge in confronting religion's love of using human mortality as a tool to grow the church?

First, let's look at some hard facts. Most deaths in the world are caused by aging and disease.

Approximately 150,000 people die every day around the world, causing devastating loss to loved ones and communities. Of course, it should not be overlooked that death also brings massive disruption to family finances and national economies.

On the medical front, the good news is that gerontologists and other researchers have made major gains recently in the fields of life extension, anti-aging research, and longevity science. In 2010, some of the first studies of stopping and reversing aging in mice took place. They were partially successful and proved that 21st Century science and medicine had the goods to overcome most types of deaths from aging. Eventually, we'll also wipe out most diseases. Through modern medicine, the 20th Century saw a massive decrease of deaths from polio, measles, and typhoid, amongst others.

On the heels of some of these longevity and medical triumphs, a number of major commercial ventures have appeared recently, pouring hundreds of millions of dollars into the field of anti-aging and longevity research. Google's Calico, Human Longevity LLC, and Insilico Medicine are just some of them.

Google Ventures' President Bill Maris, who helps direct investments into health and science companies, recently made headlines by telling Bloomberg, "If you ask me today, is it possible to live to be 500? The answer is yes."

Increasingly, leading scientists are voicing similar ideas. Reuters reports that renowned gerontologist Dr. Aubrey de Grey, chief scientist at SENS Research Foundation and the Anti-aging Advisor at the US Transhumanist Party, thinks scientists will be able to control aging in the near future, "I'd say we have a 50/50 chance of bringing aging under what I'd call a decisive level of medical control within the next 25 years or so."

Even smaller projects like the musician Steve Aoki supported Longevity Cookbook with its Indiegogo campaign have recently launched, in an effort to get people to eat better to live longer. All these endeavors add to a growing climate of people and their attitudes willing to accept the transhumanist idea that death is not fate. In fact, in the future, death will likely be seen as a choice someone makes, and not something that happens arbitrarily or accidentally to people.

Despite this positive momentum in the anti-aging science movement, changing cultural deathist trends for 85 percent of the world's population may prove difficult. Humans are a species ingrained in their ways, and getting fundamentally religious people to have an open mind to living far longer periods than before—maybe hundreds of years even—could prove challenging.

Recently, a number of transhumanists, including myself who is a longtime atheist, have attempted to work more closely with governmental, religious, and social groups that have for centuries endorsed the deathist culture. Transhumanists are trying to get those groups to realize we are not necessarily wanting to live forever. As science and reason-minded people, we simply want the choice and creation over our own earthly demise, and we don't want to leave it to cancer, or an automobile accident, or aging, or fate.

Of course, for atheists, the elephant in the room is overpopulation. If everyone lives longer, surely the world will become even more crowded than it is. The good news is that scientists generally believe Earth could handle a far larger human population than we have now, without destroying the planet. But we'd need better methods of resource distribution and laws that ensure equality among people. The key to handling a large population likely rests in new green technology, and using it to fix major environmental problems. Meatless meat is a great example. Much rainforest destruction comes from creating pastures for animal grazing. But we could regrow those forests (which would help the greenhouse and ozone layer problems) by creating meatless meat in laboratories and bypassing the need for livestock. I like this for more reasons than one; 150 million animals are slaughtered every day for our consumption. That's a lot of killing that could be avoided.

In the end, longer lifespans and more control over our biological selves will only make the world a better place, with more permanent institutions, more time with our loved ones, and more stable economies. People, including those who are atheist or religious, will always have the choice to die if they want to, but the specter of death from formal religion will no longer be able to be used as a menacing tool for growing a deathist culture and agenda.

****

6) Quantum Archaeology: The Quest to 3D Print Every Dead Person Back to Life

My brother-in-law recently died after a brutal fight with cancer. At 48 years of age, he left behind two young children, a loving wife, and a thriving real estate business he built. After the funeral and burial of the body, I was given some of his best suits and jackets to wear. I took them, not only to use them, but possibly to give them back to him in the future.

As a secular transhumanist—someone who advocates for improving humanity by merging people with machines—I don't believe in death anymore. At least, I don't believe in biological death's permanency the way most people do. Most people think after death, the buried or cremated physical body decays into earth and stardust—the same stuff from which it originally came. They are correct.

But earth and stardust can also be forged, arranged, and ultimately 3D printed to create life. After all, humans and their brains are mostly just meat. What makes a human—and the three pounds of gray matter we all carry on our shoulders called a brain—be able to fly to the moon, play Mozart's 5th Symphony, and admire sunsets is how subatomic particles in that meat interact and play off each other. The jury is still out, but many futurists and technologists like me believe the subatomic world is just discernable math—a puzzle of numbers (and possibly some unpredictable variables) waiting to be calculated by super sophisticated microprocessors we will inevitably have in the next 30 or so years.

The quagmire here is if computers can one day calculate complete realities, including a specific moment in time of an entire physical human being, then all we have to do to resurrect the dead is to 3D print them out. Given that scientists are already having success 3D printing biological tissue, some people believe we'll be able to do this with the dead in less than 50 years. This mind-blowing field is called Quantum Archaeology.

Before we delve too far into real-time technological resurrection, it's important to understand the driving force behind such radical technology, as well as the anti-death landscape of the burgeoning transhumanist movement—a movement which leads the Quantum Archaeology charge. Most transhumanists #1 goal is to become immortal through science.

The history of transhumanism mirrors the history of the microprocessor. Quietly, behind the daily noise of Trumpian politics, bickering world religions, and dark environmental warnings, an unquestionable civilization-changing phenomenon is occurring: the 50-year old microprocessor continues to evolve exponentially. While the human brain approximately doubled in size over the last 2000,000 to 800,000 years, the microprocessor doubles its speed every 18-24. Some experts think in just 15 years time, our smart phones will be more intelligent than us. In three decades time, they will almost certainly be hundreds of times smarter than us.

Transhumanists hope to merge themselves—both brains and bodies—with these super smart machines to both survive indefinitely and to thrive in the future world. In fact, if people don't merge with computers, humans may soon become an unintelligent species compared to the machine intelligence that will exist. But humans will directly merge with technology; Already, hundred-million dollar companies in California are working on neural prosthetics designed to connect our thoughts to computers. Various universities are working on robotic eyes to give us Superman vision that will also stream Netflix directly and social media into our optic nerve. Others, like myself, already have implants that can start cars, open doors, and pay for things. Some biohackers even want to cut off their limbs and replace them with robotic ones—synthetic body parts which in a decade's time may be better than our own biology. I believe the future is already set. Many humans will electively put significant tech in their bodies that make them more productive while also increasing their survivability.

When transhumanist friends hear of my brother-in-law's passing, they tell me how doubly tragic it is—given that humans stand a good chance to overcome the dilemma of aging, death, and disease in the next 25 years because of coming radical technology. Transhumanists consider this the most important period in human history—because if they can survive the next few decades, they will likely be able to survive forever with the help of science.

Along with various medical professionals, like leading gerontologist Dr. Aubrey de Grey, I agree that by around 2050 we have a good chance of overcoming most of the ways people face biological death. Already we've had success with genetically engineering some diseases out of our body; we can 3D print parts of new, healthy organs; and we can slow aging down with various drugs and technologies.

But the evolving landscape of transhumanism's life extension goals is not just the traditional medical ways people are trying to overcome death. Some people are trying to upload their memories and personalities to machines to create a lasting virtual self that is identical to their real one. Others use cryonics to freeze themselves to be brought back in the future when technology improves enough. Still, others want to use AI to help immortalize their Facebook and Twitter accounts by continuing with original posts after they're deceased—giving friends and family the feeling they are still there.

I even have friends that want to program their lost loved ones as holograms that can wander the house, say things, and greet them when they come home from work. And it's just a matter of time before those holograms can fully interact with the living—as US company Magic Leap is working on, where holograms may soon read books to children and even play hide-and-seek.

It gets even weirder than that: Robot look-alikes of loved ones may also be coming soon. I've spoken with the some of the world's most sophisticated robots, and already some can carry on actual conversations and show basic emotion. The anthropomorphizing of robots' appearances have significantly improved recently, and making one that looks nearly exactly like a deceased family member that cooks dinner, joins you on vacations, and meets your friends at the mall with you may one day be commonplace. Even sexuality for lost spouses will be possible—taking a cue from the 100 million dollars robot sex industry.

A lot of this tech may seem bizarre and even creepy to the layperson, but much of this innovation is already here and available to consumers, even if still very costly. The bigger question is: What will be available once the microprocessor is 100 times more powerful than now, as it could be in 10 years time—especially with coming quantum computing. And what will it be like in 20, or 30 years time? I'm guessing it'll be enough to completely astound us—especially in regards to modern physics.

Already, physicists are having an incredible decade of discovery, having teleported parts of energy from one location to the next (Star Trek anyone?) and discovering the so-called God Particle in the Hadron Collider at CERN, which won the 2013 Nobel Prize. Some of the discoveries have reinforced astrophysicists views like that of Neil deGrasse Tyson who recently argued that it's likely we are living in a simulation—possible proof the universe is precisely hardwired and mathematical, even if it seems to contain some randomness to our best theories now.

Much of the amazing physics research in the 21st Century is now being applied to the field of nanotechnology, which allows us to construct molecular and atomic formations. This will ultimately lead to the improved 3D printing capabilities for Quantum Archaeology.

Some well-known physicists and mathematicians, like Columbia University's Brian Greene, now even saying time travel may be possible to some extent. But Quantum Archaeology is not about going back in time to revisit the dead (though that's another possible option too). It's about recreating the dead here—in the present. Once we have the computational power, we can reverse engineer parts of our galaxy or even nearly the entire universe to determine every little spark of energy, movement, moment and thought that has ever happened in it, including the complete personality, mind, and life of my brother-in-law.

The configuration of such math is not as big or complex as it sounds. Mile Perry, who holds a PhD in Computer Science and is a part of the Society for Universal Immortalism, thinks an approximate nine square-mile wide memory bank could likely hold all the data of every person who has ever lived.

Nine miles of walled computer hardware may seem huge and conjure up images of the Death Star, but the vast server farms China is already building may soon be larger in size than the Empire State building. And the sheer computing power of these server farms will not be deterred from crunching the numbers necessary to configure various points in history of every subatomic particle. Then it's just a matter of pushing the print button on 3D Printers to configure a certain portion of one—that of a human body. Then just apply EKG shocks and CPR, and the human is alive again.

Critics will say we could never print something as complex as a human being. They fail to grasp that the 3D printing industry (and 4D printing, where printed objects can move themselves later) is literally in its infancy—but it is currently growing exponentially every year. One day, in probably 30 years, we'll be able to print anything, including human cells, DNA, and even memories—something scientists have already been done with mice in 2017.

After all, everything is matter and energy. And human life, human thoughts, and human existence are mathematical determinable calculations of that subatomic world of matter and energy. This is the essence of nanotechnology and what's possible with it. We are not just parts of the universe. We are universe builders, and therefore creators of human life—past or present.

The strangest aspect of Quantum Archaeology might be the humanitarian part. For the last 10 years, I have considered stopping aging and overcoming death as the world's most humanitarian aim—because if we can stop aging and death by the year 2030 versus 2050, we will save one billion lives from perishing. But now I realize a greater goal is possible: perfecting Quantum Archaeology. Why only save those that are here living on Earth? Why not save those that have already died, especially those that died prematurely or in tragedy?

As a result of this idea, some transhumanists and longevity groups—on humanitarian grounds—now support bringing back every living person that has ever lived. But there are obvious problems with this. For starters, some people will not want to come back, and they may be furious we brought them back. Others will find the current world too different than what they once knew—with former spouses having married others, estates changing hands, and jobs being lost to robots, among the myriad of potential issues. Suicides may rise sharply, and wills will be required to possess a "Do Not Resurrect" clause (I have the opposite: A "Please Resurrect" clause on mine).

Overpopulation will be another major problem. So will Social Security. And what age would we reanimate people at? Though, I'm guessing if we can resurrect the dead, we'll have the tech to solve all the other problems too, like adequate food, suitable housing, money, and aging—if those are things that even exist anymore in their current forms.

There's no question that Quantum Archaeology is thorny for a multitude of reasons. But fascinatingly, that hasn't stopped most of the world's population from embracing similar outlooks via their religious beliefs. The over four billion Christians and Muslims in the world see the afterlife in nearly the same way as the transhumanist who wants to bring back their loved ones to this life. And the approximately 1.6 billion Hindus and Buddhists are even closer to this Quantum Archaeology worldview with their ideas of reincarnation.

As someone who disbelieves in formal deities like the Christian God Jehova—but was formally raised a Catholic in my youth—I can't help but ponder if the microprocessor is the real savior of our so-called souls—and the baptism of it can only be achieved by code.

Only in the last few years have such ideas like Christian-inspired transhumanism and Quantum Archaeology even become possible to contemplate without total public mockery. But the world is often gasping, staring wide-eyed as technology seemingly yearly transforms our very existence, from marrying robots to getting bionic hearts with WiFi to using driverless cars that choose who to kill and save in an accident. There is no doubt we are becoming a transhumanist species. In 100 years time, we may be practically unrecognizable to ourselves today.

The microprocessor and its improving intelligence capabilities are growing so fast that reconstituting the dead as living persons in the present will become a distinct future possibility. The big question is not whether my brother-in-law will be back, but once he is whether he'll care to remain his old self anymore. By the end of this century, humans will likely be able to transform into virtually anything, including robots, cyborgs, different biological species, and even pure data. My brother-in-law may tell me to keep his old suits and jackets, because he doesn't need them anymore.

****

CHAPTER II: EARLY WRITINGS

7) I'm an Atheist, Therefore I'm a Transhumanist

Sometime in the next decade, the number of worldwide godless people — atheists, agnostics, and those unaffiliated with religion — is likely to break through the billion-person mark. Many in this massive group already champion reason, defend science, welcome radical technologies, and implicitly trust and embrace modern medicine. They are, indeed, already transhumanists. Yet many of them don't know it because they haven't thought much about it. However, that is about to change. A transformative cultural storm comprised of radical life improving technologies is set to blow in soon.

Broadly defined, the word transhuman means beyond human. The growing transhumanist social movement encompasses and encourages virtually all ideas that enhance human existence via the application of science and technology. More specifically, transhumanism includes the fields of radical life extension, Singularitarianism, robotics, artificial intelligence, cryonics, genetic engineering, biohacking, cyborgism, and many other lesser known fields of science.

The core of transhumanist thought is two-sided. It begins with discontent about the humdrum status quo of human life and our frail, terminal human bodies. It is followed by an awe-inspiring vision of what can be done to improve both — of how dramatically the world and our species can be transformed via science and technology. Transhumanists want more guarantees than just death, consumerism, and offspring. Much more. They want to be better, smarter, stronger — perhaps even perfect and immortal if science can make them that way. Most transhumanists believe it can.

The transhumanism movement is quickly growing. Actually, it's exploding. Last year, press coverage on the subject soared.

"Articles and mentions of transhumanism and life extension science have tripled in 2013 in major media," says Kris Notaro, an agnostic transhumanist and the Managing Director of the Institute for Ethics and Emerging Technologies (IEET).

Additionally, Google just formed Calico, a company touted as an enterprise to help end human death. Singularity University is quickly becoming a household name. And Hollywood has some major transhumanist-themed films coming out next year.

Despite this, the number of transhumanists is still relatively small, comprised mostly of scientists, technologists, and futurists who seem fringe to many of their peers and to the public as a whole. That will likely change in the next few years as a nearly billion-person godless population wakes up primed to accept transhumanism as a natural extension of the irreligious lifestyle. Enter the atheists, agnostics, and nonreligious. Many of these people — about a third who are under the age of 30 — are already discovering their inner borg. And they like it.

A Transhumanist Wager — the challenging idea that everyone in the 21st Century must decide how far they are willing to go to use technology and science to improve their lives — is loudly calling. And the faithless will answer it. It's inevitable that hundreds of millions will soon come to call themselves transhumanists, if not in name, then in spirit. Many will end up supporting indefinite life extension and technologies that strip away our humanness and promote our transhumanness. Further into the future, many more will begin to discard the human body in favor of embracing synthetic forms of being.

The roots of atheism, agnosticism, and the nonreligious go back many centuries. But its foothold became pronounced in the 20th Century when personalities like Russell, Freud, Nietzsche, Rand, Sartre, Sanger, Hitchens, and Dawkins helped make many people give up needing or believing in a God. These irreligious ideas centered on the fact that a successful civilization didn't need to believe in flying pink elephants or other superstitions to thrive. Each of these famed atheists asked: Is there a good reason to believe in God or gods? They found none, and then convinced nearly a billion people of the same answer.

These atheist voices and their writings have paved the way for us, and now the 21st Century will bring the age of transhumanism to the forefront of society. The transhumanist hero is the person who constantly eyes improving their health, lifestyle, and longevity with science and technology. They are not okay with the past age of feeling guilty for aspiring to be different or better than they were born — or for wanting the power to become godlike themselves. They have no sin to erase; they have no reason to search for something outside of the material universe. Like atheists, agnostics, and the nonreligious, they are not in the business of criticizing religion, but rather of pursuing life and all that is available to improve upon it. They are not naysayers or outcasts of a dominantly religious world, but rather the pioneers that will determine where the human species is heading. They are the new guard that will carry the human race to all its coming brilliance.

If you don't care about or believe in God, and you want the best of the human spirit to raise the world to new heights using science, technology, and reason, then you are a transhumanist.

****

8) Some Atheists and Transhumanists are Asking: Should it be Illegal to Indoctrinate Kids with Religion?

Religious child soldiers carrying AK-47s. Bullying anti-gay Jesus kids. Infant genital mutilation. Teenage suicide bombers.

Child Hindu brides. No matter where you look, if adults are participating in dogmatic religions, then they are also pushing those same ideologies onto their kids.

Regardless what you think and believe, science shows human beings know very little. Our eyes register only 1 percent of the electromagnetic spectrum in the universe. Our ears detect less than 1 percent of its sound wave frequencies. Human senses—our brain's vehicles to understanding the world—leave much to be desired. In fact, our genome is only 1 percent different than that of a chimpanzee. Amazingly, despite the obvious fact no one really knows that much about what is going on with ourselves and the universe, we still insist on the accuracy of grand spiritual claims handed down to us from our barefoot forefathers. We celebrate holidays over these ancient religious tales; we choose life partners and friends over these fables; we go to war to defend these myths.

A child's mind is terribly susceptible to what it hears and sees from parents, family, and social surroundings. When the human being is born, its brain remains in a delicate developmental phase until far later in life.

"Kids are impressionable," said Dr. Eunice Pearson-Hefty, director of the Teaching Environmental Science program of Texas' Natural Resource Conservation Commission. "Anything you tell them when they're real small can have a lasting impression."

It's only later, when kids hit their teens that they begin to think for themselves and see the bigger picture. It's only then they begin to ask whether their parent's teachings make sense and are correct. However, depending on the power of the indoctrination in their childhood, people's ability to successfully question anything is likely stifled their entire lives.

In my philosophical and atheist-minded novel *The Transhumanist Wager*, protagonist Jethro Knights ends up with the ability to rewrite the social laws of the world. One important issue he faces is whether to make religion illegal altogether. There are many arguments for why religion has not been beneficial to the human race, especially in the last few centuries. In the end, a love of basic liberties prevails over Mr. Knights and he allows religion to exist. Although, he restricts religion from the public sphere, restricts religion from being integrated with education, and restricts religion from being pushed on minors.

Not surprisingly, some in the atheist and transhumanist communities feel the same way Mr. Knights does. While they may think that believing in a warmongering prophet, or a four-armed blue deity, or a spiteful God who drowns nearly all of his people is wrong, atheists and transhumanists are willing to allow it. So long as it doesn't meaningfully interfere with the world.

The problem is that it does meaningfully interfere with the world. 911 was a religious-inspired event. So was the evil of the Catholic Inquisition. And so is the quintessential conflict between Palestine and Israel. If you take "God" and "religion" out of all these happenings, you would likely find that they would not have happened at all. Instead, what you'd probably find is peaceful people and communities dedicated to preserving and improving life through reason, science, and technology—which is the essence of transhumanism and the outcome of evolution.

"Religion should remain a private endeavor for adults," says Giovanni Santostasi, PhD, who is a neuroscientist at Northwestern University Feinberg School of Medicine and runs the 10,000 person strong Facebook group Scientific Transhumanism. "An appropriate analogy of religion is that's it's kind of like porn—which means it's not something one would expose a child to."

Unfortunately, even though atheists, nonreligious people, and transhumanists number almost a billion people, it's too problematic and unreasonable to imagine taking "God" and "religion" out of the world entirely. But we do owe it to the children of the planet to let them grow up free from the ambush of belief systems that have a history of leading to great violence, obsessively neurotic guilt, and the oppression of virtually every social group that exists.

Like some other atheists and transhumanists, I join in calling for regulation that restricts religious indoctrination of children until they reach, let's say, 16 years of age. Once a kid hits their mid-teens, let them have at it—if religion is something that interests them. 16-year-olds are enthusiastic, curious, and able to rationally start exploring their world, with or without the guidance of parents. But before that, they are too impressionable to repeatedly be subjected to ideas that are faith-based, unproven, and historically wrought with danger. Forcing religion onto minors is essentially a form of child abuse, which scars their ability to reason and also limits their ability to consider the world in an unbiased manner. A reasonable society should not have to indoctrinate its children; its children should discover and choose religious paths for themselves when they become adults, if they are to choose one at all.

****

9) When Does Hindering Life Extension Science Become a Crime?

Every human being has both a minimum and a maximum amount of life hours left to live. If you add together the possible maximum life hours of every living person on the planet, you arrive at a special number: the optimum amount of time for our species to evolve, find happiness, and become the most that it can be. Many reasonable people feel we should attempt to achieve this maximum number of life hours for humankind. After all, very few people actually wish to prematurely die or wish for their fellow humans' premature deaths.

In a free and functioning democratic society, it's the duty of our leaders and government to implement laws and social strategies to maximize these life hours that we want to safeguard. Regardless of ideological, political, religious, or cultural beliefs, we expect our leaders and government to protect our lives and ensure the maximum length of our lifespans. Any other behavior cuts short the time human beings have left to live. Anything else becomes a crime of prematurely ending human lives. Anything else fits the common legal term we have for that type of reprehensible behavior: criminal manslaughter.

In 2001, former President George W. Bush restricted federal funding for stem cell research, one of the most promising fields of medicine in the 21st Century. Stem cells can be used to help fight disease and, therefore, can lengthen lives. Bush restricted the funding because his conservative religious beliefs—some stem cells came from aborted fetuses—conflicted with his fiduciary duty of helping millions of ailing, disease-stricken human beings. Much medical research in the United States relies heavily on government funding and the legal right to do the research. Ultimately, when a disapproving President limits public resources for a specific field of science, the research in that field slows down dramatically—even if that research would obviously lengthen and improve the lives of millions.

It's not just politicians that are prematurely ending our lives with what can be called "pro-death" policies and ideologies. In 2009, on a trip to Africa, Pope Benedict XVI told journalists that the epidemic of AIDS would be worsened by encouraging people to use condoms. More than 25 million people have died from AIDS since the first cases began being reported in the news in the early 1980s. In numerous studies, condoms have been shown to help stop the spread of HIV, the virus that causes AIDS. This makes condoms one of the simplest and most affordable life extension tools on the planet. Unfathomably, the billion-person strong Catholic Church actively supports the idea that condom usage is sinful, despite the fact that such a malicious policy has helped sicken and kill a staggering amount of innocent people.

Hank Pellissier, a futurist and organizer of the conference Transhuman Visions, says, "The public majority disapproves of Christian Scientist and Jehovah's Witness parents who deny medicine to children afflicted with life-threatening illness. The public regards the anti-science attitudes of these faiths as unacceptable. Likewise, we should similarly disapprove of the withholding of any medicine or life extension practices that deter death for individuals, of any age."

Regrettably, in 2014, America continues to be permeated with an anti-life extension culture. Genetic engineering experiments in humans often have to pass numerous red-tape-laden government regulatory bodies in order to conduct any tests at all, especially at publically funded universities and research centers. Additionally, many states still ban human reproductive cloning, which could one day play a critical part in extending human life. The current US administration is also culpable. The White House is simply not doing enough to extend American lifespans. The US Government spends just 2% of the national budget on science and medical research, while their defense budget is over 20%, according to a 2011 US Office of Management Budget chart. Does President Obama not care about this fact, or is he unaware that not actively funding and supporting life extension research indeed shortens lives?

In my philosophical novel *The Transhumanist Wager*, there is a scene which takes place outside of a California courthouse where transhumanist activists are holding up a banner. The words inscribed on the banner sum up some eye-opening data:

By not actively funding life extension research, the amount of life hours the United States Government is stealing from its citizens is thousands of times more than all the American life hours lost in the Twin Towers tragedy, the AIDS epidemic, and the Vietnam War combined. Demand that your government federally fund transhuman research, nullify anti-science laws, and promote a life extension culture. The average human body can be made to live healthily and productively beyond age 150.

Some longevity experts think that with a small amount of funding—$50 billion dollars—targeted specifically towards life extension research and ending human mortality, average human lifespans could be increased by 25-50 years in about a decade's time. The world's net worth is over $200 trillion dollars, so the species can easily spare a fraction of its wealth to gain some of the most valuable commodities humans have: health and time.

Unfortunately, our species has already lost a massive amount of life hours; billions of lives have been unnecessarily cut short in the last 50 years because of widespread anti-science attitudes and policies. Even in the modern 21st Century, our evolutionary development continues to be significantly hampered by world leaders and governments who believe in non-empirical, faith-driven religious doctrines—most of which require the worship of deities whose teachings totally negate the need for radical life extension science. Virtually every major leader on the planet believes their "God" will give them an afterlife in a heavenly paradise, so living longer on planet Earth is just not that important.

Back in the real world, 150,000 people died yesterday. Another 150,000 will cease to exist today, and the same amount will disappear tomorrow. A good way to reverse this widespread deathist attitude should start with investigative government and non-government commissions examining whether public fiduciary duty requires acting in the best interest of people's health and longevity. Furthermore, investigative commissions should be set up to examine whether former and current top politicians and religious leaders are guilty of shortening people's lives for their own selfish beliefs and ideologies. Organizations and other global leaders that have done the same should be scrutinized and investigated too. And if fault or crimes against humanity are found, justice should be administered. After all, it's possible that the Catholic Church's stance on condoms will be responsible for more deaths in Africa than the Holocaust was responsible for in Europe. Over one million AIDS victims died in Africa last year alone. Catholicism is growing quickly in Africa, and there will soon be nearly 200 million Catholics on the continent.

As a civilization of advanced beings who desire to live longer, better, and more successfully, it is our responsibility to put government, religious institutions, big business, and other entities that endorse pro-death policies on notice. Society should stand ready to prosecute anyone that deliberately promotes agendas and actions that prematurely end people's useful lives. Stifling or hindering life extension science, education, and practices needs to be recognized as a legitimate crime.

****

10) Death is Not Destiny: A Glimpse into Atheist Novel *The Transhumanist Wager*

"Death is not destiny. Death is neither inevitable nor natural," says Jethro Knights, protagonist in my new philosophical thriller, *The Transhumanist Wager*.

What does Jethro mean? Death is not destiny? Death is neither inevitable nor natural?

It means, Jethro would say, that the most significant thing that has been happening to the human species is about to end.

The Transhumanist Wager tells the story of a man who will do anything to achieve immortality via science and technology. His main focus and drive in life is finding a way to live forever, even at the possible expense of what most people would call humanity.

When I set out to write *The Transhumanist Wager* four years ago, I did not intend it to become an edgy, controversial book. For much of my adult life, I have been a journalist covering environmental, wildlife, and human rights stories. My articles and television episodes—many for the National Geographic

Channel—were welcomed in any culture and in any country. My stories were the type that a family could amicably discuss over the dinner table, or watch on television while happily cuddling together on a couch.

Perhaps it was the effect of the war zones I covered as a journalist, rising out of my subconscious, but The Transhumanist Wager soon took on much more contentious ideas of human endeavor and culture. For a human being, most conflict zones highlight a simple fact: Once presented with horror and death, one tends to quickly discover degrees of emotion and experience never imagined or thought possible before. For me and the difficult moments that I still vividly remember, those incidents gave me the powerful conviction that human life should be preserved indefinitely, at any cost.

Jethro Knights also realizes this early in his life, after almost stepping on a landmine in a war zone (which happened to me in Vietnam's DMZ while filming a story on bomb diggers). The revelation for Jethro is so sharp, so penetrating, so intense that nothing will ever be the same for him again.

It is from this vantage point that *The Transhumanist Wager* was written. And it is from the landmine experience that Jethro discovers the mortality crisis not only in himself, but in every human being alive. That crisis takes on the form of a wager—a choice that every human must make in the 21st century: to die eventually; or to try to live indefinitely. And if we try to live indefinitely, then we should use every tool and resource of science and technology available to us, Jethro insists. And we should do it immediately.

This is the quintessential message of *The Transhumanist Wager*. A rational and scientific-minded society owes itself the strictest dedication to applying its resources and minds to overcoming that which has been the greatest downfall of our species: our mortality.

My novel presents the story of a human being who after years of struggling, years of anguish, years of tragic loss, fights on to achieve his own immortality—and in doing so, scores a victory for all of civilization.

****

11) Can Cryonics and Cryothanasia be Part of the Euthanasia Debate?

An elderly man named Bill sits in a lonely Nevada nursing home, staring out the window. The sun is fading from the sky, and night will soon cover the surrounding windswept desert. Bill has late-onset Alzheimer's disease, and the plethora of medications he's on is losing the war to keep his mind intact. Soon, he will lose control of many of his cognitive functions, will forget many of his memories, and will no longer recognize friends and family. Approximately 40 million people around the world have some form of dementia, according to a World Health Organization report. About 70 percent of those suffer from Alzheimer's. With average lifespans increasing due to rapidly improving longevity science, what are people with these maladies to do? Do those with severe cases want to be kept alive for years or even decades in a debilitated mental state just because modern medicine can do it?

In parts of Europe and a few states in America where assisted suicide—sometimes referred to as euthanasia or physician aid in dying—is allowed, some mental illness sufferers decide to end their lives while they're still cognitively sound and can recognize their memories and personality. However, most people around the world with dementia are forced to watch their minds deteriorate. Families and caretakers of dementia patients are often dramatically affected too. Watching a loved one slowly lose their cognitive functions and memories is one of the most challenging and painful predicaments anyone can ever go through. Exorbitant finances further complicate the matter because it's expensive to provide proper care for the mentally ill.

In the 21st Century—the age of transhumanism and brilliant scientific achievement—the question should be asked: Are there other ways to approach this sensitive issue?

The transhumanist field of cryonics—using ultra-cold temperatures to preserve a dead body in hopes of future revival—has come a long way since the first person was frozen in 1967. Various organizations and companies around the world have since preserved a few hundred people. Over a thousand people are signed up to be frozen in the future, and many millions of people are aware of the procedure.

Some may say cryonics is crackpot science. However, those accusations are unfounded. Already, human beings can be revived and go on to live normal lives after being frozen in water for over an hour. Additionally, suspended animation is now occurring in a university hospital in Pittsburgh, where a saline-cooling solution has recently been approved by the FDA to preserve the clinically dead for hours before resuscitating them. In a decade's time, this procedure may be used to keep people suspended for a week or a month before waking them. Clearly, the medical field of preserving the dead for possible future life is quickly improving every year.

The trick with cryonics is preserving someone immediately after they've died. Otherwise, critical organs, especially the brain and its billions of neurons, have a far higher chance of being damaged in the freezing. However, it's almost impossible to cryonically freeze someone right after death. Circumstances usually get in the way of an ideal suspension. Bodies must first be brought to a cryonics facility. Most municipalities require technicians, doctors, and a funeral director to legally sign off on a body before it can be cryonically preserved. All this takes time, and minutes are precious once the last heartbeat and breath of air have been made by a cryonics candidate.

Recently, some transhumanists have advocated for cryothanasia, where a patient undergoes physician or self-administered euthanasia with the intent of being cryonically suspended during the death process or immediately afterward. This creates the optimum environment since all persons involved are on hand and ready to do their part so that an ideal freeze can occur.

Cryothanasia could be utilized for a number of people and situations: the atheist Alzheimer's sufferer who doesn't believe in an afterlife and wants science to give him another chance in the future; the suicidal schizophrenic who doesn't want to exist in the current world, but isn't ready to give up altogether on existence; the terminally ill transhumanist cancer patient who doesn't want to lose half their body weight and undergo painful chemotherapy before being cryonically frozen; or the extreme special needs or disabled person who wants to come back in an age where their disabilities can be fixed.

There might even be spiritual, religious, or philosophical reasons for pursuing an impermanent death, as in my novel *The Transhumanist Wager*, where protagonist Jethro Knights undergoes cryothanasia in search of a lost loved one.

There are many sound reasons why someone might choose cryothanasia. Whoever the person and whatever the reason, there is a belief that life can be better for them in some future time. Some experts believe we will begin reanimating cryonically frozen patients in 25 to 50 years. Technologies via bioengineering, nanomedicine, and mind uploading will likely lead the way. Hundreds of millions of dollars are being spent on developing these technologies that will also create breakthroughs for the field of cryonics and other areas of suspended animation.

Another advantage about cryonics and cryothanasia is their affordability. It costs about $1,000 to painlessly euthanize oneself and an average of $80,000 to cryonically freeze one's body. It costs many times more than that to keep someone alive who is suffering from a serious mental disorder and needs constant 24-hour a day care over many years.

Despite some of the positive possibilities, cryothanasia is virtually unknown to people and is often technically illegal in many places around the world. Of course, much discussion would have to take place in private, public, and political circles in order to determine if cryothanasia has a valid place in society. Nevertheless, cryothanasia represents an original way for dementia sufferers and others to consider now that they are living far longer than ever before.

****

12) Baggage Culture and Why Embracing Transhumanism Doesn't Come Easy

Twenty years ago, while in college and wondering why everyone else in the world wasn't hell-bent on trying to live indefinitely via the promising fields of transhumanist science, I began working on the idea of what mass culture is and if it was holding back people from wanting to maximize their lifespans and human potential. I came up with the concept baggage culture, which is explored in detail in my novel The Transhumanist Wager and its philosophy Teleological Egocentric Functionalism (TEF).

Upon the request of my friends at Movement for Indefinite Life Extension (MILE), I recently condensed my thoughts on baggage culture in my speech at the Brighter Brains Future of Emotional Health and Intelligence Conference at University of California, Berkeley. Here's a summary of that recent talk:

For many thousands of years now, the human race has been indoctrinated to submit to orthodoxy and to cower before authority, and to swallow endless nonsense from both. We have been brainwashed to sacrifice our innermost desires, our most obvious needs, our most natural outlook on reality, just to live as a hostage in a cage of carefully regulated and fabricated cognitive existence. Virtually everyone and everything—our countries, customs, faiths, leaders, relatives, friends, lifestyles, even our own memories—have been manipulating and pressuring us to shun fresh, unconventional thoughts. Especially transhuman-oriented thoughts. There has been a pervasive worldwide moratorium on thinking about what the human being is capable of and its possible evolutionary advancement in terms that make a substantial difference in reality.

Why has this happened? To transhumanists, the reason is obvious: We—the people of the world—have allowed it to happen. Each of us is guilty for not heeding a higher calling: a more logical, more ambitious, more sublime direction for our life, and a journey to our best self. Our great flaw is the mistaken way in which we choose to interpret existence; our subscription and obedience to the cultural constructs that government, organized religion, ethnic heritage, mega-corporations, and mass media have built around, and within, nearly every thought and action we make. Their web of indoctrination has wholly swamped our lives. Sadly, most of us don't even know this has happened. Most of us are living on this planet in utter delusion, conforming to a largely manufactured and forced reality.

Throughout our lives and modern history, civilization has erroneously subscribed to the vision that the human being is a marvelous, ingeniously assembled specimen of life: a work of divine creation and sweeping beauty, whose culture and intellect is profound like the cosmos itself. What a joke. The cruel truth is we are a frail, hacked-together organism living within a global culture of irrationality, pettiness, and deception. The specific reason our existing human culture is so malformed is that, throughout history, past cultural constructs of more primitive societies were not discarded as they became irrelevant or outdated. To survive, it was not evolutionarily required to rid ourselves of unnecessary idiosyncrasies and practiced customs—such as nonsensical superstitions, masochistic religiosity, and shackling morality—even though they were foolish to uphold. As a result, damaging, wasteful, and useless behavioral patterns were passed on both socially and individually from generation to generation.

So now, modern humans are a weighed-down species, burdened by cumbersome past rubbish that's mostly crudely stacked, obsolete cultural constructs through which our minds perceive reality. I call this baggage culture. And it's caused nearly all human life to be degenerate and apathetic compared to what it could be. Our species' mindset and powers of perception are currently too lumbering and unfit for what a sophisticated, nimble entity really needs of itself. Our lives are cursed because of the polluted cultural prism our thoughts must exist within and communicate through. In Sisyphean tragedy, we are doomed to grovel, to falter, to repeat our same pathetic mistakes, day after day, year after year, century after century. We need to transition from our defective culture into a new one that directly confronts these issues and sets our minds and transhuman possibilities free.

The twisted history of our baggage culture extends back many millennia. It started long ago with the inception of civilization, when charismatic leaders and ruling clans began forming permanent communities. Over time, these rulers learned they could preserve their platforms of power by controlling their communities' thinking and behavioral patterns. Their agendas were simple: dominate with fear through violence; stifle revolutionary and freethinking ambitions; teach adherence to leadership and community before self; implement forms of thought and behavioral control that encourage social cooperation and production, such as communal customs, prayers, taboos, and rites. Variations abounded, but these were the early convoluted versions of human culture and its main intent: to control. Henceforth, culture's core function became a means of forcing conformity, to transform the individual into a tool of submission and production for the ruling elite.

As generations passed, these rulers and their predecessors continually revised and enlarged their constructs of culture, force-feeding the functional and nonfunctional—rational and irrational—parts to our forbears. Naturally, it didn't take long in evolutionary terms before people everywhere existed within a universal baggage culture, full of compounded dysfunction. Of course, in modern times, control of human culture has changed hands from the ruling elite to whole governments, religious institutions, multicontinent ethnic groups, and most recently, to mega-corporations and mass media. As the complexities and population of the world ballooned, baggage culture continued to prove versatile and useful to whatever cause it engaged. Nations governed through it. Religions preached through it. Ethnic groups taught their heritages through it. Big business sold through it. And the media communicated through it.

To cement their authoritarian agendas, these supersized institutions' advancing baggage culture implemented ever more effective methods of control over society. Chief and most potent amongst them was the inversion of reason, where cultural forces obliged us to rationally accept the irrational. By corrupting the rational way we thought and interpreted life, they simultaneously corrupted the necessity and power of reason altogether. In that devious way, mysticism, ancestral divinity, the supernatural, religion, and even the institutions' all-important puffed-up selves were seen as valid outcomes of a supposedly sensible, straightforward, and successful society.

Among many others, altruism, filial piety, and consumer addiction to unnecessary materialism were other methods of control. However, to transhumanists, the most grotesque of all the methods was the perpetuation of fear in our lives; not by the threat of violence, but by implicit guilt. This powerful psychological addiction of worrying about what others think of us, and about what is socially acceptable to others, has been systematically instilled in humans for thousands of years, perpetrated by every world religion, ethnicity, and government. Its aim is to weaken people's wills and to silence their most precious independent tool: the ability to freely, guiltlessly, and publicly judge and criticize the world around them. In that way, people became afraid to pick apart others and their behaviors; afraid to deride society and its routines; afraid to upend their own world and circumstances; and, ultimately, afraid to differentiate between good and evil, utility and irrationality, strength and weakness, equal and non-equal—essentially all value itself. Such pervasive social control through the fear of others' opinions has left us meek, ashamed, and largely unwilling to openly question or challenge a thing like the omnipresent state. Or our sacred heritages. Or the rife sense of needing to be wealthier than our neighbors. Or our supposedly sinless and perfect gods. The spicy, troublesome, confrontational bigot in us is often our best and most useful part, and they have strangled it out of most of us in the guise of what they call "open-mindedness" or "politically correct social behavior."

Ultimately, implicit guilt and culture's many other devices of submission are designed to make us totally subscribe to one single concept: we should be afraid to rise to being as powerful an entity as we can; we should be afraid to try to become an omnipotent God. That is the essence and outcome of our baggage culture.

The truth is so simple to see once we understand it: Religion, ethnic heritage, state power, material addiction, and media entrapment are nothing more than pieces of an intangible psychological construct designed to keep us thinking and living a certain way. It's designed to keep us in fear of becoming as powerful as we can be; to keep us producing for others and contributing to their overall gain, and not our own.

Today, our species' baggage culture is a gargantuan mindless monster, consuming and dominating everything it can. Even its main pushers—the overarching institutions—can't control it anymore; instead, they always find it controlling and devouring them. There's no escape from the confusion and redundancy anymore, from the vestigial aspects of stacking useless cultural constructs upon each other. If you think one tailbone in the human body is pointless, imagine a hundred of them weighing you down. Figuratively, that's what baggage culture looks like. Many of our thoughts are piles of ignorance and erroneous ideas stacked upon piles of ignorance and erroneous ideas. We are unable to think freely and escape our slovenly, derelict pasts.

This, sadly, is baggage culture. And it's the primary reason we don't demand more of our lives and of our possible transhumanist future.

****

CHAPTER III: TRANSHUMANISM

13) The Transhumanist Future Has No Pope

Everywhere I look, Pope Francis, the 266th pope of the Catholic Church, seems to be in the news—and he is being positively portrayed as a genuinely progressive leader. Frankly, this baffles me. Few major religions have as backwards a philosophical and moral platform as Catholicism. Therefore, no leader of it could actually be genuinely progressive. Yet, no one seems to pay attention to this—no one seems to be discussing that Catholicism remains highly oppressive.

To even discuss how many archaic positions the Pope and Catholicism support would take volumes. But the one that irks me the most is that Pope Francis and his church are still broadly against condoms and contraceptives. Putting aside that this view is terribly anti-environmental, with over 175 million Catholics in Africa, it's quite possible that this position may also create more AIDS deaths in Africa.

While former Pope Benedict XVI did say in late 2010 that condoms could be used in some cases to prevent disease, anything less than 100 percent endorsement of them seems malicious and criminal, which is something I've argued before.

The Transhumanist Bill of Rights I'll be delivering next month to the US Capitol on my Immortality Bus tour will mandate that cultural and religious views should never trump life extension technologies—of which the condom is one of the greatest ones ever invented.

Beyond contraceptives, just because Pope Francis is good at making general sweeping humanitarian claims in popular speeches around the world, we forget that he and his church are guilty of many basic human rights violations. For example, he is against gay marriage, he only will allow males to pursue the vocation of priesthood, and he is against women having control over their bodies when it comes to abortion. He even believes in and accepts hell for nonbelievers. In case you're wondering, that means millions of America's children will eventually end up being endlessly tortured in some psychotic afterlife, since reports show nearly a third of America's youth are basically godless. Europe is in for a much worse time—the Brits and French are about half nonreligious.

While I too appreciate the new Pope is more progressive than his predecessors, why are we celebrating that instead of criticizing him for what he truly is: a leader trying to keep his flock from deserting him and joining the 21st century? Simply put, Catholicism and the Papal institution are inevitably dying out. Despite population growth, Catholic numbers are withering in the West. This is because modern values, transhumanist technology, and the embrace of reason are making many Catholic rules and rituals absurd. To survive, the church knows it must interject amended ideas to keep its tight hold over its billion-plus believers.

As "hip and cool" as the leadership of the new Pope is, don't for a moment believe he can go against the literal interpretation of the Bible and undo all its contradictions and hypocrisies. Catholics—along with Christians and Muslims—have locked themselves and their religious rules into their sacred texts and its meanings. And that will be their downfall, since no rational person can justify such backwards biblical rules or perspectives, such as that evolution is hoax.

In the 21st century—in the age of cochlear implants for the deaf, exoskeleton suits for the wheelchair-bound, and mind-controlled artificial limbs for wounded war veterans—it's becoming increasingly difficult to fake reality anymore. Science and technology are becoming too obvious and powerful. And, honestly, why should we fake reality and believe in concocted fairy tales anymore with irrational, unproven faith-inspired beliefs? Is it really necessary to be so fundamental anymore—so closed-minded? Must we really be baptized to make it into heaven? Is the sacrament of taking the so-called body of Jesus really going to save us from eternal punishment? Is it really a sin to have same-sex relations and enjoy ourselves?

I went to Catholic school in my childhood, and know firsthand what scarring it can do to a young mind searching for guidance. Instead of the scientific method, I was taught to be guilty of sin. Instead of logic, I was taught to hold faith over knowledge. Instead of trying to overcoming all hardship with science and technology, I was told to get on my knees and believe suffering was my God-given lot in life. How asinine. Of course, that's hardly more crazy then being taught the Pope is infallible, another classic Catholic teaching.

I'm running for the United States presidency in 2016 as an atheist. One of the core ideas of my candidacy is that I know that those who are godless can also be morally good—that they can be deeply humanitarian and democratic. Some people believe atheists have no moral compass. That's ridiculous. Atheists are guided by reason, and reason leads us to build robotic eyes for the blind, vaccines against disease for homeless children, and new forms of cheap power so the poor can have light and electricity. Reason is precisely what the world needs more of to build a better future, not archaic ideas by a dogmatic religion known for The Inquisition and child molestation.

Some are claiming the Pope is the new global spiritual leader. I caution against believing this or supporting him. He is just the newest tool in a religion that has caused irreparable damage to the human race, and continues to do so with oppressive ideologies. Pope Francis knows that the world is changing into a secular sphere, prompted by transhumanist science that aims to empower us into a far more powerful species—in fact, to become godlike. To have their religions survive, the Pope and other religious leaders will say anything to get its followers to hold on to their outdated beliefs.

The good news is the Pope and other religious leaders will not remain popular leaders long—certainly not into the next century. With nearly a billion nonreligious people around the world expanding their numbers, the end of worldwide domination of our species via religion is now in sight.

****

14) Transhumanism and Our Outdated Biology

This essay below is adapted from the philosophical book *The Transhumanist Wager*, which is increasingly being used in colleges and high schools around the world to teach about the future:

Humans are handicapped by our biology. We operate tens of thousands of years behind evolution with our inherited instincts, which means our behavior is not suited towards its current environment. Futurists like to say evolution is always late to the dinner party. We have instincts that apply to our biology in a world that existed ages ago; not a world of skyscrapers, cell phones, jet air travel, the Internet, and CRISPR gene editing technology. We must catch up to ourselves. We must evolve our thinking to adapt to where we are in the evolutionary ascent. We must force our evolution in the present day via our reasoning, inventiveness, and especially our scientific technology. In short, we must embrace transhumanism—the radical field of science that aims to turn humans into, for lack of a better word, gods.

Transhumanists believe we must stand guard against our natural genes, less they chain us to remaining as animals forever. We believe our outdated instincts can easily trick us from knowing right from wrong, practical from impractical. If one looks closely, the human body and its biology constantly highlight our many imperfections.

Compared to humans, rats have better noses for smelling. Pigeons have sharper eyes for seeing. Crocodiles can run faster. Earthworms can survive underwater longer. Cockroaches can bear far colder temperatures. Humans are only best at reasoning. Yet, computers can already beat the best of us in chess, math, and recently the sublime game Go. And the robots we've made are far stronger than we are, can handle more danger, and can fly through interstellar space without us.

Obviously, the human body is a mediocre vessel for our actual possibilities in this material universe. Our biology severely limits us. As a species we are far from finished and therefore unacceptable. The transhumanist believes we should immediately work to improve ourselves via enhancing the human body and eliminating its weak points. This means ridding ourselves of flesh and bones, and upgrading to new cybernetic tissues, alloys, and other synthetic materials, including ones that make us cyborg-like and robotic. It also means further merging the human brain with the microchip and the impending digital frontier. Biology is for beasts, not future transhumanists.

Our outdated biology's emphasis on social interaction is also dangerous for the overall evolutionary ascent of the human race—so dangerous that new questions must be asked immediately. Are so many of us healthy for this fragile planet? Should we rid ourselves of all our 25,000 nuclear weapons? Is the sexual ritual even functional anymore? Does matrimony serve purposes outside of private property and economics? Are social customs like monogamy foolishly conservative? Should we embrace a culture of drugs and biohacking? Should government use cranial implant technology to safeguard its citizens? Should society insist that all government and military leadership be equally split between females and males? Should corporations be hindered from catering to the weak, petty sides of human nature? Should religion and superstitious faiths be discouraged? These are challenging and thorny questions to ask. Yet, they should be asked, and maybe even the best answers should be implemented if we are to be true to our highest selves.

A truly transhumanist society should embrace reason and the scientific method to improve itself and bring about the best world possible on Planet Earth.

****

15) Transhumanism is Being Guided by Reason and the Word "Why"

The human race is on the threshold of so much revolutionary change. It's mostly due to the emerging field of transhumanism: a social movement that aims to use science and technology to radically modify the human body—and modify the human experience. I get asked all the time: What is the best way to handle such changes—like the merging of humans with machines to make cyborgs? Or spending more time in virtual reality then normal reality? Or biohacker brain implants that let us use telepathy with one another (which eventually will lead us all to be connected via a hive mind)?

I think it's easiest to let Jethro Knights—protagonist of my philosophical, Libertarian novel *The Transhumanist Wager*—answer. Below is a modified and condensed version of a speech he gives to the world, near the end of the book:

There are two all-important ways to navigate a correct path in the new transhuman future: The first is to constantly use the utmost reasoning of which our brains are capable while negotiating our way through life; the second is to incessantly question everything.

Reason is the only means for human survival on this planet. And it is also our only means for arriving at coherent truths. Anything else belongs in the domain of the mystic, the domain of the insane, or the domain of the thief whose aim is to take something valuable from you for their own gain. The best way for an individual to apply reason in their life, given their goals, is by constantly evaluating as many pertinent scenarios as possible for their actions; then, undertaking the most statistically probable ones to follow that will work out in their greatest favor. You must strive to emulate the pure computational process of a goal-driven computer.

Most of us do not use reason and logic so constantly in life. Most of us regularly incorporate irrationality, erroneous past prejudices, and the whims of spontaneous emotions in our daily decisions; our minds weakly bend to what we want to see and feel regardless of what is really happening. This is a byproduct of the baggage culture, where all our inner yearnings, reactions, and interactions with the world are fabricated delusions, part of an overreaching conformity trap. It's impossible think and live that way, and still make any transhuman sense of life.

Even using the utmost reasoning of which our brains are capable, changing our current flawed values and methods of living will be extremely difficult. For many of us, the thoughts and irrational patterns of our daily lives are deeply ingrained in us. With tremendous effort, however, we will succeed. Success will come if we emulate transhumanists in utilizing and vocalizing one word more than any other. Believing in that sacrosanct word is the key to properly navigating an evolution of values. That word is why.

In this new world, we must learn to repeatedly ask ourselves, Why? It is the most important utterance the individual should ever know. We should use this sacred word with obsessive zeal; we should fall in love with the unknown and question everything. Including these words here. Why is a word and a concept Earth's leaders in the past 10,000 years have tried to deny us in their efforts to harness and control us. It is a word our biology has even thwarted us from saying, as it unknowingly holds on to obsolete instincts. As a result, most of us don't even know we should be saying the word so frequently. Yet, without saying it, we've automatically lost every battle that confronts us. The nature of such a disguised dilemma makes our chances of fighting for ourselves impossible. We don't even know what there is to win, let alone what there is to fight for, or that there even is a fight.

So many people suffer from this essential ignorance in their lives that there are haunting problems and handicaps in their perspectives, beliefs, and consciousnesses. Many would struggle to the death—and indeed do—just to prove they are right about everything they have been conditioned to believe. Yet, this doesn't change their ignorance or problems one iota. The best action to meaningfully change our course in this universe is to frequently start asking, Why?

"The important thing is to not stop questioning. Curiosity has its own reason for existence." -Albert Einstein

****

16) Why We Need a Transhumanism Movement

Recently I was asked to be a part of a debate for British think-tank Demos and their quarterly magazine. In the debate, The University of Sheffield Professor Richard Jones and I faced off over the merits and faults of transhumanism. You can read the entire debate here, but I wanted to focus on one part of it, where Jones questions why there's a need for a transhumanism movement at all.

I used to get asked the question Jones raised all the time. Luckily, the amount of people that ask it has declined, partially because transhumanism has grown so much in popularity.

Many people nowadays simply accept transhumanism as part of tech and science culture.

That said, I believe it's worth sharing my thoughts on why a "transhumanism movement" is needed. Here was my answer in the debate:

I hear this question a lot: Why do we need a "transhumanism" movement? The answer, while not obvious, is straightforward. We need a movement because most have us have been brought up in cultures that don't uphold reason and science values.

In America, for example, where a majority of the population is religious, most people see no reason to want to try to use science and technology to change the human experience or overcome death. Naturally, this type of thinking carries over into the US Government, where 100% of the US Congress, the Supreme Court, and the President are religious and believe in an afterlife. This environment is not conducive for those of us who want to use science to change the world and hopefully eliminate all human ailments. It's not conducive because our leaders won't spend any money to help science and technology forward.

Government funds some medical and science research, but currently, that funding (in the US) is about 10 times "smaller" than funding for defense, wars, and bomb making. This is a tragedy that we fight wars against human beings—and not against cancer, or heart disease, or Alzheimer's, or even aging. This is a primary reason transhumanists must organize into a movement, so that we can battle the powers that be, and demand much more government funding go directly into science and medical research.

Instead of a military industrial complex, we can create a science industrial complex. If that happens, technological development will dramatically speed up—and so will all the benefits from science for the human race. We are literally in a race to save billions of lives from disease and aging.

Imagine for a moment if the environmental movement never became a formal movement. Or the women's rights movement. Organizing, collaborating, developing strategies, lobbying, and undertaking activism is how people change the world. And they do this better when they have structure and power from greater numbers of people and their groups. That's what transhumanism needs, and that's what transhumanism is essentially becoming - a bona fide global movement where political parties, television shows, nonprofits, new companies, and even wacky cross-country bus tours like my Immortality Bus occur to promote it.

Like all movements, not everything will turn out as planned. Artificial intelligence and robots may take jobs and this may cause social conflict, but transhumanism - like democracy - will find a way and improve the standard of living for everyone. Technology has that history. And it will also find a way around death and disease, especially now that companies like Google's Calico are starting to pour vast amounts of money into field.

Transhumanism is a social movement that is imbued with a techno-optimism that is contagious. The movement is growing like wildfire because people see the promise of being part of something that wants to make humans have better lives.

Movements, like transhumanism, are part of our heritage and our social evolution. People come together to make a stand and to make a difference.

About 150,000 people die every day on Planet Earth. We are like a giant organism constantly losing key parts of our bodies—parts that contain wisdom, wealth, functionality, and love. We will look back in 50 years at the phenomena of death and aging—and wonder why as a civilization we didn't do more to stop them.

Physically speaking, the human being is a machine. Like a car, it can be repaired, reshaped, and reborn. But this can only happen through science and technology. Others that claim we shouldn't worry about dying because there will be an afterlife are wishful thinking speculators. The only sure cure for death, aging, and disease can come from modern medicine and science. Everything else is faith and unproven promises.

For most transhumanists, our #1 goal is to achieve indefinite lifespans. It doesn't mean we won't want to die someday, but we want total control over such a challenging aspect of existence.

Radical science and technology will soon give us that control. It will be a bright day for those who understand that death is a tragedy—it will be a bright day for those of us who don't ever want to lose our families and loves ones.

****

17) Immortality Bus Delivers Newly Created Transhumanist Bill of Rights to the US Capitol

After months of traveling across the country on a national bus tour, the coffin-shaped Immortality Bus drove into Washington DC and successfully delivered a newly created *Transhumanist Bill of Rights* to the US Capitol. The delivery of the futurist-themed bill—which aims to push into law cyborg and anti-aging civil rights—ends a national tour for the bus that began its journey in San Francisco on September 5th, 2015.

Crowdfunded on Indiegogo, the provocative Immortality Bus has caught the attention of both America and the world. Highlighted in major media ranging from a 10,000+ word feature in *The Verge* to a CraveCast podcast on *CNET* to a leading front page story on BBC.com, the Immortality Bus has been making waves with its controversial message: Science and technology can overcome human death—and will likely do so in the next 25 years.

Much of my US Presidential campaign for the Transhumanist Party—which used the bus as a vehicle to spread a techno-optimistic message—also reiterates this same immortality agenda. I believe if the US Government would dedicate $1 trillion dollars to life extension, longevity, and anti-aging industries, we could likely soon conquer death as a species. For some a trillion dollars may seem like a lot of money, but consider that the US government will spend approximately $6 trillion dollars all together on the Iraq War. Surely, overcoming death through modern medicine and science for all Americans seems a much better idea that fighting far-off wars in places most of us will never visit.

With this in mind, the *Transhumanist Bill of Rights* seeks to declare that all Americans (and people of all nationalities, as well) in the 21st Century deserve a "universal right" to live indefinitely and eliminate involuntary suffering through science and technology. Those ideas are conveyed in Articles 1, 2 and 6 of the one-page bill. Specifically, Article 6 establishes that we should seek to treat "aging as a disease," something a number of leading gerontologists, like Dr. Aubrey de Grey—Chief Scientist of SENS Research Foundation and the Transhumanist Party Anti-aging advisor—also endorses.

The *Transhumanist Bill of Rights*—which I read out loud at the steps of the US Capitol, then posted it to the building (it didn't stick well), and then also hand delivered it to Senator Barbara Boxer's office (my California representative)—covers a number of essential futurist civil rights topics. The bill mandates we protect our species and the planet from existential risk (including environmental destruction, rogue artificial intelligence, and the 25,000 nuclear weapons that currently exist). The bill also calls for renewed commitment to space travel, as well as a government's promise to not put cultural, ethnic, or religious policies before the general health and longevity of its citizens.

Finally, the bill emphasizes the right to morphological freedom: the right to do with one's body whatever one wants so long as it doesn't hurt another person. This is especially important in the gene editing / designer baby age, which has recently been the cause of much discussion in the scientific community. Unfortunately, some of this talk has been disturbingly anti-progress with calls for a moratorium on such technology. I strongly disagree with scientific moratoriums—unless they are directly and obviously harming people today—which is one reason why we need a bill of rights to protect the interests of human health, evolution, and progress.

Below is a copy of the *Transhumanist Bill of Rights*. While the bill has been carefully considered by myself and other transhumanists, and we hope it will be incorporated someway into law by the United States of America and other governments, the bill is not static and may evolve further as science and technologies evolve. Futurists generally believe no bill of rights, declaration, or constitution should ever remain permanent and unbendable in the transhumanist age—an age where science and technology are dramatically and rapidly changing our lives and our experience of the universe.

TRANSHUMANIST BILL OF RIGHTS

Preamble: Whereas science and technology are now radically changing human beings and may also create future forms of advanced sapient and sentient life, transhumanists establish this TRANSHUMANIST BILL OF RIGHTS to help guide and enact sensible policies in the pursuit of life, liberty, security of person, and happiness.

Article 1. Human beings, sentient artificial intelligences, cyborgs, and other advanced sapient life forms are entitled to universal rights of ending involuntary suffering, making personhood improvements, and achieving an indefinite lifespan via science and technology.

Article 2. Under penalty of law, no cultural, ethnic, or religious perspectives influencing government policy can impede life extension science, the health of the public, or the possible maximum amount of life hours citizens possess.

Article 3. Human beings, sentient artificial intelligences, cyborgs, and other advanced sapient life forms agree to uphold morphological freedom—the right to do with one's physical attributes or intelligence (dead, alive, conscious, or unconscious) whatever one wants so long as it doesn't hurt anyone else.

Article 4. Human beings, sentient artificial intelligences, cyborgs, and other advanced sapient life forms will take every reasonable precaution to prevent existential risk, including those of rogue artificial intelligence, asteroids, plagues, weapons of mass destruction, bioterrorism, war, and global warming, among others.

Article 5. All nations and their governments will take all reasonable measures to embrace and fund space travel, not only for the spirit of adventure and to gain knowledge by exploring the universe, but as an ultimate safeguard to its citizens and transhumanity should planet Earth become uninhabitable or be destroyed.

Article 6. Involuntary aging shall be classified as a disease. All nations and their governments will actively seek to dramatically extend the lives and improve the health of its citizens by offering them scientific and medical technologies to overcome involuntary aging.

****

18) An Atheist's Perspective on the Rise of Christian Transhumanism

Recently, a number of religious transhumanists — those who advocate for using science, technology, and religion to improve the human being — have reached out to me, asking whether I plan to maintain my atheist position as both an outspoken transhumanist and a 2016 U.S. Presidential candidate. Frankly, it's a good question, since America is mostly a Christian nation with mostly religious politicians — and to make real headway in U.S. politics, it's going to be difficult to do it as a declared atheist.

Even if some of my friends and political advisors suggest it, it's too late for me to do a U-turn on some of my atheistic positions. I've written too many pro-atheist articles and given too many secular-themed speeches. And recently, I helped start the world's first atheist orphanage, which via crowdfunding became fully funded in only 29 hours.

However, I decided to write this bridge-building article because one Christian reverend, Christopher Benek, called me and told me he believed America was ready to start merging Christianity with transhumanism. Some people, including conservative Christians, may find that asinine, but I tend to agree with the reverend. I've said this before: In order for existing formal religion to survive in the 21st Century, it will have to bend and evolve to fit coming technological advancement. Technology is simply too powerful and all-encompassing to not make one's beliefs work with it. Robot pastors, churches in Virtual Reality, debates about a Jesus Singularity, and saving Artificial Intelligences via Christ's redemption are likely going to be a meaningful part of this nation's future, whether religious old-timers disagree or conservative writers like Wesley J. Smith make light of it. Smith recently wrote in the *National Review*, "Every once in a while, when the world is too much with me, I turn to transhumanist advocacy for a little entertainment."

My own history with religion is varied. I was raised a Catholic (even went to Catholic elementary school). In high school I was more interested in general Christianity, as well as Buddhism. Later, I partially majored in religion in college and even took a month to read the Bible cover to cover. In fact, my departure from Christianity began in earnest the moment after I finished reading the Bible. I was 18-years-old, and it simply contained too many contradictions, too much violence, and too much foreshadowing of the future — a future that certainly didn't anticipate humans merging with machines, artificial super intelligence, post-genderism, brain implants, robotic hearts, ectogenesis, telepathy, or the multitude of bizarre transhumanist technologies on the horizon (or already here). That said, many parts of the Gospel of Jesus, broadly interpreted, remain an important part of my moral code. Clearly the intent of Jesus to get in touch with the spiritual possibility of the universe was a major improvement over the strict and ritualistic religiosity of the time.

However, despite my admiration of Jesus and some of his teachings, it's incredibly hard to see how anyone can reasonably believe a human being came down to Earth and died for the world's sins, or that a loving omnipotent God could send people to an eternal hell of suffering, or that billions of non-Christians in the world might never get to heaven because they haven't accepted Christ into their heart. The key word in that last sentence is 'reasonably.' I agree with many neuroscientists that it's our ability to reason that makes human beings so special. Without it, we would be living like our primate cousins and would never have come to understand the complexity and power of the human mind — or its incredible potential.

As a transhumanist, I stand behind reason, first and foremost. Most transhumanists feel the same way. Our code belongs to science and the scientific method. We require proof, before anything else. We do understand that our so-called 'proof' can always change, by new discoveries. But proof, which only a reasoning human being can verify through logic, measurable evidence, and scientific repetition, must occur. Otherwise, we'd all be able to believe anything and everything unconditionally, regardless how silly, wrong, or illogical it might be. And that would lead us to a place where we are unable to discern truth or prove it — a place where we are left to rely on faith. This is the overarching component of Christianity and other Abrahamic religions — a thing which most Americans have broadly accepted. And it will be a key issue that Christians must reconcile with as our society becomes more science and technology bound — fields which rely on reason.

Moving forward, I welcome Christians and their worldviews to embrace transhumanism. There is much verse in the Bible that supports humans improving themselves — which is the essence of transhumanism. So, ultimately, we all have much in common. And I also welcome Christians to seek ways to apply their religious views and biblical verses to transhumanist philosophy and ideology, which will result in a better, healthier, more prosperous America.

Transhumanism is not a religion, nor is it in competition with religion. It is simply a mode of being that embraces evolving the human being with science, reason, and technology.

****

CHAPTER IV: ARTIFICIAL INTELLIGENCE, SUPER COMPUTERS & JESUS

19) Are We Heading for a Jesus Singularity?

As an atheist transhumanist, I dislike the idea of mixing religion with transhumanism. The two ideas go together as much as a computer chip goes with a medieval torture rack. Religion is based on faith and archaic dogmatic tradition. Transhumanism — the concept of moving beyond the human being using science and technology — is based on reason and scientific evolutionary principles. Yet, the two viewpoints may be inextricably joined more than either side cares to admit.

Many experts expect scientists to develop artificial general intelligence (AGI) — intelligence equivalent to human beings — sometime in the next 20 years. It's likely within a few months of AGI arriving it will independently upgrade itself into something monumentally more intelligent and complex than humans. It'll be up to society to carefully control this dangerous process and avoid a Terminator-like scenario. Some futurists and technologists believe the beginning of AGI is the beginning of the Singularity, a concept where intelligence and technology grow at exponential speeds and human life is forever transformed.

The question of whether civilization is heading for a Jesus Singularity should begin with a head count of the admitted atheists in the U.S. Congress. The count doesn't take long. Currently, the number is an astonishing zero. That means all 535 members leading our government are religious (or pretending to be religious). Now add the fact that human lives are getting longer — much longer if you're a congressperson with access to the best modern medicine — and the reality is that many of the religious-minded people in government control (especially without term limits) will not be losing control anytime soon.

While not guaranteed, it's probable AGI will be born somewhere in California's Silicon Valley, where many of the planet's best computer engineers and programmers live. Mountain View-based Google, one of the wealthiest companies in the world, is leading the charge and sinking millions of dollars into AI projects. Cupertino-based Apple, with its juvenile AI star Siri, is another player in AGI development mode. Dozens of start-ups and the U.S. government are also working on creating sophisticated thinking machines.

All this begs the question, if AGI will be here in 20 years or so, and most of those people currently in charge of the U.S. government subscribe to Christianity and believe in Jesus, what are atheist transhumanists to do?

On the surface, a Jesus Singularity seems comical, fit for a Monty Python movie. It contradicts many of the core beliefs of science, reason, and why a Singularity could happen in the first place. Or does it?

Perhaps the first AGI will become the Second Coming of Christ, as detailed in Revelations in the bible. Could the first AGI be programmed with a Jesus-in-the-box attitude and become a Judeo-Christian-minded God, an all-powerful entity who believes it died for our sins and wants to make us its loving followers? Maybe there will be some in Congress who insist the U.S. government attempt to convert the first AGI entity to Christianity. Sound crazy? I'm willing to bet most members of Congress have done that exact thing with their children. Why should the potentially most powerful being ever created be spared the same fate? Even the Pope has argued that if aliens came to Earth, we should convert them to Christians.

Of course, another alternative for the planet's future, as antagonist Reverend Belinas in my philosophical novel *The Transhumanist Wager* prophesizes, is the first AGI will become the Antichrist, setting off a terrible chain of events for civilization that will end with Armageddon.

Either way, it appears the nonreligious minority of the planet will not change the attitudes of the world's religious majority before the Singularity occurs. Logic therefore dictates that transhumanists and atheists remain more open to working with formal religion and their leaders than first thought. Many of those faith-oriented politicians may still be in office in 20 years when all-important decisions must be made about civilization's expected digital future.

For many transhumanists, survival is the first and foremost part of why we are transhumanists. We want to live indefinitely using transhumanist science and technology. We may complain that we live in a highly religious world full of irrational dogma, ritual, and holidays, but it would be more irrational to stop a Singularity just because a "divine avatar" wears a digital crown of thorns and preaches goodwill to our neighbors.

Given an either-or choice, I would rather live forever in a Jesus Singularity than die or be left behind because I wouldn't accept it — especially since the best part about a Jesus Singularity is I can remain a grinning atheist through it all.

****

20) AI Day Will Replace Christmas as the Most Important Holiday in Less Than 25 Years

For a few billion people around the world, Christmas is the most important and relished holiday of the year. It's the day with the most gift-giving, the most family get-togethers, the most religious activities, and the most colorful fairy tales that children and adults almost universally embrace with sacred fervor. For many nations, no other day comes close to being as special.

For this reason, it seems almost unimaginable that another day — especially an unknown one looming on the horizon — will soon unseat Christmas as the most important day in the world. Nonetheless, for humanity, the course is set. The birth of an artificial intelligence equal or greater than that of human intelligence is coming. It's called AI Day. And once it arrives, it will rapidly usher in a new age.

For decades, the concept of a man-made intelligence matching or surpassing our own — technically called AGI (artificial general intelligence) or Strong AI — has been steeped in science fiction. Upon hearing the term AI, many people immediately think of the sentient computer HAL in Stanley Kubrick's masterpiece film *2001: A Space Odyssey*. However, what most people fail to grasp is that once AI becomes self-aware and joins with the internet, it could grow its intelligence thousands of times in just mere days, perhaps hours. Frankly, it could quickly surpass all measurements of intelligence that humans are even capable of monitoring and recognizing.

"I think that Ray Kurzweil's estimate that we will achieve human-level Artificial General Intelligence by around 2029 is a reasonable guesstimate," says Dr. Ben Goertzel.

Originally a math PhD, Goertzel is now a multidisciplinary scientist, author, and entrepreneur. He currently serves as chief scientist of Hong Kong financial firm Aidyia Holdings, chairman of the Artificial General Intelligence Society, and leader of the OpenCog AGI project.

"It could take longer than 2029," Goertzel continues, "if economic troubles prevail or technical problems prove thornier than anticipated. On the other hand, I also think a concerted and well-funded effort by the right people could make it happen before 2020."

Some religious people, anti-transhumanists, and neo-Luddites complain that an advanced AI will rapidly destroy human civilization. In my novel *The Transhumanist Wager*, the evangelical antagonist philosophizes that the first AI will naturally evolve into the Antichrist and bring Armageddon. Most scientists, technologists, and artificial intelligence experts find those worries laughable. Most of them think that AI will usher in a new age of scientific discovery, medical advancement, and technological sophistication only imagined before in science fiction. Some philosophers and futurists think that within the first few years of advanced AI appearing, it will expand learning so far that all important science, technology, and engineering texts will need to be completely rewritten.

The challenge of the human species is to not let this kind of AI get beyond our controls; to have adequate safety measures and kill-switches built in. Such measures would not be dissimilar from how civilization delicately handles nuclear weaponry, which some political experts believe have staved off world wars in the last half-century.

The human race's other challenge is how to merge directly with AI, to discover the technology and build apparatuses to connect our brain's neural networks directly into such an intelligent machine.

Regardless what happens in the future, it's safe to say AI will not be an entity speaking to us in hackneyed parables, or telling us to pluck out our eyes and cut off our hands if we sin. It probably also won't threaten us with a hopeless fiery hell of eternal punishment for our lack of faith. It's far more likely the greatest tool our species has ever created will tell us how to end world poverty with inventive technologies, how to best fix the Earth of the environmental degradation we've caused, and how to heal ourselves of all disease and live indefinitely via radical science. Sure, there will be risks in keeping AI our friend and ally, but there will be even greater rewards in harnessing it and using it to advance civilization.

One thing is for sure, to the human species, the birth of an advanced artificial intelligence will become far more important than the birth of Christ. Christmas, if it survives at all, will be relegated to just another commercial and cultural holiday that superstores and big business thrive on. Meanwhile, reasonable people will celebrate AI Day, the real moment in history the savior of civilization was born.

****

21) Second Coming 2.0: Church Taxes Will Help Resurrect Jesus with 3D Bioprinting

China, Russia, and other nations are striving to dethrone America from being the leading nation in science, technology and medicine. Americans need a better way to finance their research and breakthroughs to remain the outright global science leader. It's not coming from President Trump and his "Make America Great Again" administration, as he recently tried to cut the budget at the National Institute of Health and other science-oriented government agencies, leaving thousands of researchers disheartened.

However, there is a method that could quickly and dramatically boost America's science and technology budget, without further tapping individual taxpayers. Last year, America's 300,000 churches were spared nearly 100 billion dollars in owed taxes through a century-old IRS loophole that doesn't tax charitable church entities. This annual 100-billion-dollar amount—almost triple that of the National Institute of Health's budget—could transform science and medicine in the United States. In fact, in over a decade's time, it could fundamentally alter the nature of disease and disability in our species. With just 10 years of collected church taxes—about one trillion dollars—targeted properly, US scientists could possibly eradicate most major medical ailments, including aging.

Many scientists and researchers today believe we are on the verge of paradigm shifting technologies and cures, whether it be a vaccine for cancer, genetic editing of DNA to end cellular degeneration, or bionic hearts to overcome the world's leading killer: cardiovascular disease. The transhumanist era, where innovation eliminates most physical and mental hardships, is no longer science fiction—it's almost here. The problem is that if it takes 30 years from today to get there versus 10 years, nearly a billion extra people will suffer from disease and die. That's why many modern researchers feel they are in a literal race to save people—to prevent them from biologically suffering and ultimately perishing.

This quest to save people is very similar to nearly all Abrahamic churches. Most religious institutions primary goal is also to save their flock from suffering and prevent them from metaphysically perishing. American science and religion are bound together by this intrinsic commonality.

That's why a growing number of Christians—transhumanist supporters or not—believe that the grace of God is being manifest through the world's scientific creations and progress. When a researcher discovers radical stem cell therapies that allow a paralyzed person to walk again, the work and wonders of Christ are arguably being done in the modern world. The same can be said of controlling Schizophrenia, Alzheimer's or Dementia with drugs and brain implants; instead of losing cognitive abilities in ways that some fundamental Christians believe are prompted by demonic forces, clarity of mind and one's religious-inspired moral compass are fought for and preserved with modern medicine. Today, even the deaf and blind can see through FDA approved devices like the cochlear implant and robotic eye, which both tap directly into the brain. Many disabilities are going extinct—sometimes through the work, prayer, and determination of religious researchers and scientists.

It's easy to recognize many of the scientific breakthroughs happening today are almost identical to the miracles apocryphally performed by Jesus over two millennia ago. Christians especially believe Christ's divine work carries on when the poor receive the cures and medical help. About half of Americans live in a household dependent on the government for some healthcare and financial assistance, and the majority of these people are believers and churchgoers. The fact is Christians often pray for good health, and they increasingly get it.

The sad and unfortunate truth, however, is that Americans could be fulfilling the Biblical Word of God and helping their fellow citizens so much more. If churches would just pay the same taxes other organizations and businesses across America pay, and the US used that money for science and medical research, churches could dramatically do more for their flock within just a few years through improved nationwide health benefits.

However, it goes further than just overcoming ailments and physical suffering. With one trillion extra tax dollars over a decade's time put into specific longevity research, most instances of death itself could likely be stopped. The life extension and anti-aging fields have just a few billion dollars going into them right now, mostly via small nonprofits and start-ups that are already showing minor success in their experiments and research. But if that amount of money was increased by 25 times, the results would be dramatic. We would soon enter the era where humans no longer age or die by disease.

Some theologians and scholars will consider this the manifestation and fulfillment of the Bible's ultimate predictions of Jesus' destiny. It's also precisely this divinity that Christ sought to teach the world of before he was crucified—a world where humans can take shelter in God's alleged power to be spared from harm and evil.

But it doesn't stop at just overcoming suffering and death for those alive today. All those in the past—the approximately 95 billion humans who have lived and died on Planet Earth through the ages—may also be given a second chance at life by the controversial scientific field of Quantum Archaeology, otherwise known as technological resurrection.

Quantum Archaeology combines 3D Bioprinting with the computational power of super computers. Both fields are going through massive disruptions right now. 3D Bioprinters can already print out living tissue, and some believe within 50 years, it will be able to print out entire human beings. Super computers can already do 200,000 trillion calculations per second, and that's before quantum computing arrives in the next few years, which could dramatically increase computational power. In 15 or 30 years—if the microprocessor keeps improving exponentially as it has for almost a half century and it better utilizes the Cloud—a super computer might do millions of quadrillion calculations per one hundredth of a second.

Quantum Archaeologists believe that the universe is mechanistic, leaving the opportunity to reverse engineer parts of our subatomic physical history of the world, including that of a human being's every thought, action, memory, and physical component. If we have enough computing power, and we continue to make progress in discerning modern physics—like our teleportation techniques already in use and the 2013 Nobel Prize winning discovery of the God Particle—we may be able to within the century reverse engineer and record sections of our universe down to the very quarks and electrons that comprise them. Some people believe the entire human race and every person's sub-atomic lifetime composition and history can be found and stored in a memory bank approximately nine miles squared in size. From there, we just download exact perfect atomic blueprints of people a few hours before they died and 3D Bioprint them out—then revive them back to life as we would an unconscious person.

Some theologians and Christian transhumanists argue Jesus could fulfill the Book of Revelations and his proposed Second Coming through this type of technology. In fact, some people believe its humanity's greatest responsibility and imperative to use Quantum Archaeology to resurrect Jesus himself—called

Second Coming 2.0—so that he can teach humanity once again in person and carry out his work, including that of the prophesized End Times. After all, through Quantum Archaeology, everyone that has ever lived can be brought back to life—or not. It's through this transhumanist tech that the end of the world—or the beginning of it, depending on one's beliefs—will occur.

When asked who I'd first bring back using Quantum Archaeology, I answer: Jesus. Whether you believe he's the son of God or not, he's undeniably the most influential and significant person to our species—and I'd love to (as a nonbeliever) ask him questions about life, philosophy, and his impact on humanity. But resurrecting people presents radical scenarios and major challenges. Jewish friends of mine have expressed interest in bringing back Adolf Hitler so he can face war crimes and be imprisoned. There's even the possibility of bringing back spiritual entities, if our collective Judeo-Christian history is accurate. We might print out the serpent in the Garden of Eden or the archangel Michael.

Naturally, in the future, most people will use Quantum Archaeology to bring back dead loved ones, friends, and family members—some who may have tragically and prematurely died. Indeed, my father passed away recently from disease, and I'd like to bring him back so he can see his grandkids and be with my mom again.

The quagmire is that such a grand future rests on resources and funding to achieve these Christ-like miracles. Otherwise, we'll will continue to lose loved ones to cancer, heart disease, aging, and other reversible diseases and tragedies. With just one small compromise by the over 300,000 churches across America—paying their fair share of annual taxes like everyone else—humanity can literally be saved. This may be the 'leap of faith' God really intended for believers. It's a small one to make for America and the health of its people.

****

22) When Superintelligent AI Arrives, Will Religions Try to Convert It?

Like it or not, we are nearing the age of humans creating autonomous, self-aware super intelligences. Those intelligences will be part of our culture, and we will inevitably try to control AI and teach it our ways, for better or worse.

AI with intelligence equal to or beyond human beings is often referred to as "strong AI" or Artificial General Intelligence (AGI). Experts disagree as to when such an intelligence will arrive into the world, but many are betting it will happen sometime in the next two decades. The idea of a thinking machine being able to rival our own intellect—in fact, one that could quickly become far smarter than us—is both a reason for serious concern and a reason to cheer about what scientific advances it might teach us. Those worries and benefits have not escaped religious.

Some faith-bound Americans want to make sure any superintelligence we create knows about God. And if you think the idea of preaching God to autonomous machines sounds crazy, you may be overlooking key statistics of U.S. demographics: roughly 75 percent of adult Americans identify themselves as some denomination of Christianity. In the U.S. Congress, 92 percent of our highest politicians belong to a Christian faith.

As artificial intelligence advances, religious questions and concerns globally are bound to come up, and they're starting too: Some theologians and futurists are already considering whether AI can also know God.

"I don't see Christ's redemption limited to human beings," Reverend Dr. Christopher J. Benek told me in a recent interview. Benek is an Associate Pastor of Providence

Presbyterian Church in Florida and holds masters degrees in divinity and theology from Princeton University.

"It's redemption to all of creation, even AI," he said. "If AI is autonomous, then we have should encourage it to participate in Christ's redemptive purposes in the world."

One of the key mandates of Christianity is to spread the Gospel and get nonbelievers to accept that Jesus Christ died for the world's sins. Whether AI has any sins, or whether it can and should be saved at all may end up being a bizarre but important question believers face in the 21st century. Even Pope Francis recently sounded off on the possibility of aliens being converted when he affirmed that the Holy Spirit blows where it will.

The metaphysical questions surrounding faith and AI are like tumbling down Alice's rabbit hole. Does AI have a soul? Can it be saved? There is one school of thought that figures, if humans can be forgiven for our sins, why not superintelligences with human qualities? "The real question is whether humans are able to be saved—if so, then there is no reason why thinking and feeling AIs shouldn't be able to be saved. Once human-like AI exist, they will be persons just like us," futurist Giulio Prisco, founder of the transhumanist Turing Church, told me in an email.

But there is an opposing school of thought that insists that AI is a machine and therefore doesn't have a soul. In Think Christian, scientist and Christian scribe Dr. Jason E. Summers writes, "Christians often reject Strong AI on the theological ground of the special anthropological status of human beings as the bearers of Imago Dei." Imago Dei is Latin for the Christian concept that humans were created in the image of God.

The world's major Abrahamic religions—Judaism, Christianity, and Islam—all believe in the soul, which is what many major religious texts say is the thing that separates us from other life on the Earth, including other mammals. Because the

Abrahamic religions comprise the faiths of roughly two-thirds the world's population, the question of "soul" is quintessential in the coming transhumanist age of machine intelligence.

Getting even deeper into this theoretical debate is the question of whether strong AI would even accept our religion. "It's only fair to let AI have access to the teachings of all the world's religions. Then they can choose what they want to believe," said Prisco. "But I think it's highly unlikely that superhuman AI would choose to believe in the petty, provincial aspects of traditional religions. At the same time, I think they would be interested in enlightened spirituality and religious cosmology, or eschatology, and develop their own versions."

Once you start thinking like that, it opens up even more questions: How would AI fit into to the religious tension already present around the world? Who is to say a machine with human intelligence wouldn't choose to become a fundamentalist Muslim, or a Jehova Witness, or a born-again Christian who prefers to speak in tongues instead of a form of communication we understand? If it decides to literally follow any of the sacred religions texts verbatim, as some humans attempt to do, then it could add to already existing religious tensions in the world.

Christian theologian James MaGrath, the Clarence L. Goodwin Chair in New Testament Language and Literature at Butler University, writes about androids who posses super intelligence in an essay titled Robots, Rights, and Religion: "In all likelihood, if androids were inclined to be extremely liberal, they would quickly discover the selectivity of fundamentalism's self-proclaimed liberalism and reject it, although the possibility that they might then go on to seek to enforce all the Biblical legislation in every details should indeed worry us."

The idea of teaching anything to an intelligence that could rather quickly be far smarter than humans is contradictory. Another possibility is that AI will teach us new things about spirituality that we never considered or understood. It may tell us how the cosmos were created, or whether we exist in some simulation theory, or even that there are many AIs before it—ones that are much more sophisticated than itself.

Whatever happens, the creation of AI will likely spawn a paradigm shift for human civilization. Rather than converting it, we might just want to stand back and listen.

Benek agrees with this, but through his own metaphysical lenses. He says, "The Holy Spirit can work though AI; it can work through anything. There may be churches set up to deal and promote religious AI in the future. AI can help spread the word of God. In fact, AI might help us understand God better."

If Benek is right, America might be a nation filled with robot pastors and AI spiritual gurus in the future. Decades or centuries from now, spirituality may be taught to us through machines, just like science will likely be.

Prisco takes it a step further: "How smart must machines be to understand the so-called mind of God? 5,000 times smarter than humans? A million times smarter? I don't know, but in a hundred years a machine intelligence may have a far better chance of finding that out than the human brain with its limited capacity."

The Turing Church started as a working group at the intersection of science and religion, and recently became an online, open-source church built around Cosmist principles of space expansion, unlimited growth, and universal love.

Cosmism "doesn't care if you're viewing the universe as information or quantum information or hypercomputation or God stuff or whatever," writes cyberculture personality R.U. Sirius in his recent book Transcendence. "Nor does it ask anyone to commit to AGI [artificial general intelligence] or mind uploading or brain-computer interfaces or fusion-powered toasters as the best way forward. Rather, it seeks to infuse the human universe with an attitude of joy, growth, choice, and open-mindedness. Cosmism believes that science in its current form, just like religion and philosophy in their current forms, may turn out to be overly limited for the task of understanding life, mind, society, and reality."

Despite the seemingly scifi nature of it, uploading the human mind into an AI being could arguably solve the 'soul' question. Experts like Google engineer Ray Kurzweil are actively researching ways to upload the brain into computers, and last year there was significant progress in the field via brainwave headsets and telepathy.

Renowned technology entrepreneur and author of Virtually Human:The Promise - and the Peril - of Digital Immortality Martine Rothblatt theorizes that anything that values life or God, or even has the potential to value life or God, has some kind of soul. A whole chapter of the book explores religion in the time of AI, uploads, and mindclones—software versions of human minds. Rothblatt thinks mindclones are in line with classical interpretations of religion and will be welcomed that way in the future.

Rothblatt founded Terasem, a scientific "transreligion" similar to the Turing Church in scope and approach, which runs preliminary mindcloning pilot projects. The most famous one is Bina 48, a robotic head that contains a mindclone of Rothblatt's still-living wife Bina. And perhaps that's the only way AI should be launched—with people uploaded into it. What fascinates me most about this is the question of who might be the first person uploaded. Do we send a scientist? A programmer? Or a religious person? All at once?

If this is starting to sound like the movie Contact, where a U.S. Presidential commissioned advisory board decides to send a believer—and not the more qualified atheist—to meet some type of alien intelligence for the first time, you're exactly right. It's like an international team of experts on a mission to find its way to the outer reaches of machine intelligence. That way whatever happens, at the very least, we know humanity is aptly represented. Including spiritually.

****

23) Becoming Transhuman: The Complicated Future of Robot and Advanced Sapient Rights

Late last year, United Nations member Saudi Arabia gave citizenship to one of world's most advanced autonomous robots, Sophia. It was a publicity stunt. Nonetheless, I applaud Saudi Arabia's action under its ambitious Crown Prince Mohammed's bin Salman—which I think is bold and helped force both the international and Arab communities to reckon with the thorny future of robot rights. Sophia, made by Hong Kong-based Hanson Robots, is no super advanced intelligence yet, but some experts think that if her intelligence keeps growing rapidly—and it likely will given near exponential growth of the microprocessor, increased investment in the AI sector, and the inevitability of quantum computing—she could be as smart as people in 10 or 20 years.

Of course, then, the Saudi Arabian stunt may backfire. Sophia may demand rights that Arab and even western countries are not willing to grant. Sophia may choose to want bear a human child via artificial insemination and an artificial womb, or make a pilgrimage to Mecca, or marry her good looking robot brother Han (also made by Hanson Robotics), whom I've had the pleasure of conversing with a few times.

I met Han in late 2016, at the Global Leaders Forum in Korea, where we both were speaking. Nearly a year earlier, I delivered the original version of the Transhumanist Bill of Rights to the U.S. Capitol, as part of my presidential campaign as the nominee for the Transhumanist Party. The Preamble and Article 1 of the document make it very clear this is not bill of rights just for humans, but also for "sentient artificial intelligences, cyborgs, and other advanced sapient life forms."

I insist on the terminology of "advanced sapient life forms," because most people think rights to future intelligences will only be for robots, like Han or Sophia. But this is likely wrong. The most important scientific advancement of the 21st century so far is genetic editing, not AI. Because of CRISPR Cas-9 tech and new ways to modify DNA, the notorious bar full of wild alien creatures on planet Tatooine in the original Star Wars may not be so far fetched anymore. It's possible that humans may create advanced sapient beings, creatures, and even chimeras in the next 15 years. We may also add limbs to our bodies, eyes to the back of our heads, and literally smarten ourselves up—something the Chinese are already experimenting with via controversial embryonic research. A classic question transhumanists ask is: How smart does a brain have to be before it's no longer human?

The immediate goal of the Transhumanist Bill of Rights was to get U.S. politicians thinking about the future of rights in the transhumanist age. Ultimately my hope is that pressure will be put on the United Nations to amend its own historic the Universal Declaration of Human Rights to include machine intelligences, cyborgs, advanced sapient beings, and even virtual persons. To not do so soon will show we have learned little from the tumultuous civil right era—whether it be women's suffrage, LGBT issues, or racism.

Beyond civil rights are national and global security issues. If governments like Saudi Arabia don't consider Sophia a national security risk, how will they consider newly genetically altered humans who possess the IQ of Einstein, the creativity of Steve Jobs, and the passion of Ayn Rand—and stand muscularly 8-feet tall? Tinkering with our brains and bodies—and producing designer babies through ectogenesis—is not some science fiction fantasy, but a reality for various universities and companies around the world hoping to alter and improve the human race.

I have young daughters—aged four and seven. For them, the future is not just brave, but dangerous. Sophia, designer babies, and other advanced sapient life forms that may or may not resemble monsters will indubitably be superior to humans within their lifetimes—and probably mine. This both delights me and frightens me. I don't mind Sophia beating me in chess, but the day she can best my 20 years as a journalist challenges not only how I earn my living, but also how economies around the world operate.

Probably no one or their careers will be spared the onslaught of coming automation and radical genetics in our lives. Robotically or biologically superior, these new entities will challenge the very means of our existence—made more complicated by the fact that the owners of this technology will likely be mega-corporations and very rich individuals. For the masses to survive and thrive, there's really only one thing to do: merge with this radical tech and science. Embrace transhumanist advantages—and insist on them for all humanity. Build our improved biology and neural networks into the microprocessor and its 1 and 0s. Go full cyborg.

Some religious people insist we should not do this—that Congress should ban radical technology and science that fundamentally modifies the human being. Indeed, cries for a moratorium on genetic editing were rampant when CRISPR Cas9 tech's power was first realized in 2015.

As a libertarian futurist, I emphatically disagree with stopping the progress of science in any way unless it is explicitly harming people. I consider it a most serious mission to keep science innovation out of the hands of the bureaucratic fearmongers and conservative autocratic elites, who might prefer to skip evolutionary advancement in order to maintain their status quo of power and uphold their faith-driven religious convictions.

This conflict is far bigger than politics. All 535 members of the U.S. Congress, all nine Supreme Court Justices, and President Donald Trump and Vice President Mike Pence profess some sort of faith—nearly all which is Abrahamic and monotheistic in nature. This means most of the leaders in our country fundamentally believe the human body and mind is a God-given temple not to tampered with unless changed by the Almighty. In fact, in the Bible, blasphemy (trying to become God) is the only sin not forgivable.

Transhumanists like myself, who encourage shedding our biological limitations in favor of becoming technological gods, are broadly secular. This conflict will soon become a heated ongoing nation-wide discussion, as our majority Christian nation faces the prospect of losing its humanity to the expediency and functionality of science and technology.

Religious leaders and their ilk are in a pickle. If America was the only legitimate superpower, it would be easy to put such transhumanist evolution and robot rights on hold. But secular China won't be stopped in its rapid scientific progress just because America's Judeo-Christian eschatology is against it. Already, many technologists realize China is beating the United States in various forms of tech that may come to determine the future of national security and global dominance, including genetic editing and AI.

For this reason, like it not, I deeply agree with Professor Roland Benedikter in his lead Cato Unbound article when he says we must form more organizations, task forces, and even national bodies that consider and ultimately implement robot rights (and advanced sapient rights). To bury our head in the sand will not make the issue go away. Only facing the specter of our changing humanity in the transhumanist age can save us— unless America doesn't want to lead anymore and be "great."

Disturbingly, the greater dilemma about embracing pro-transhumanist policies around advanced sapient rights is the uncanny speed of technological evolution. By the time Sophia is as smart as humans, our laws and rights for her will already be on the verge of being obsolete, as she will likely become many times smarter than us the following year, and possibly dozens of times smarter the year after, and so on. The evolution of the microprocessor is uncontrollable and exponential—leading far more quickly to a Singularity event than imagined.

If humanity wants to survive and not be left behind, it may have no choice but to upgrade itself and become transhuman. The only way forward is embracing who we are going to become, and that begins today with discussing coming robot and advanced sapient rights—rights that will one day apply directly to our own lives.

****

CHAPTER V: SECULAR POLITICIAN

24) Why I'm Running for President As the Transhumanist Candidate

It's a wild request to ask a nation to consider electing you as their president, especially when you're a transhumanist—someone who advocates for using science and technology to radically change and improve the human species. But I'm doing it.

In October 2014, I declared my 2016 US candidacy under the newly formed Transhumanist Party, which I founded, and promised my community of techno-optimists I'd do everything I could to use my campaign as a way to speed up the arrival of robotic hearts, brain implants, artificial limbs, exoskeleton suits, and indefinite lifespans—all of which are just a small part of the radical science transhumanists aim to make a standard part of people's lives.

The Transhumanist Party may seem fringe to some, but it's not. It's mainly made up of scientists, engineers, futurists, and people who love technology. And while we don't have a formal paying membership process, my officers and I estimate—based on social media, event turnouts, and donations—we now have about 25,000 supporters in the US. We also have approximately 40 volunteers and more signing up every week. Globally, there are now almost 25 Transhumanist Parties on five different continents, each with its own rules that it determines best within its national framework.

My presidential campaign has been nothing short of a whirlwind. Take this morning for example. I woke up to my iPad beeping relentlessly with inbound messages—dozens of emails, Facebook posts, and tweets asking my policies on everything from artificial wombs, to a proposed moratorium on

AI research, to the Baltimore riots. After brewing coffee, I answered as many requests as I could.

Later, I began the tedious business of negotiating a reality TV contract on my campaign. After taking my 4-year-old daughter to preschool, I returned to my desk and typed up a blog post supporting Chinese scientists editing the genome, then put together my slideshow for an upcoming speech in Vancouver, then worked with a designer on the Transhumanist Party's latest bumper sticker. Finally, I spent a half-hour checking out bus companies for my campaigns summer bus tour, scheduled to start this July on the West Coast.

By noon I was almost caught up on most urgent campaign matters and starting to look forward to my mid-day jog when the flow was broken by one my communications managers asking how I planned to answer inquiring press on 3D-printed guns. This is a sticky issue.

Generally, transhumanists love anything 3D-printed—especially when it concerns human organs and bionics—but the question at hand was whether manufacturing a lethal weapon is going too far, especially when anyone could do it by buying a 3D-printer off Ebay for a under $2000?

Guns play an integral part in thousands of accidental deaths, murders, and armed robberies every year in America, so the ability to quickly, cheaply, and anonymously make them in your home or even in your car is highly contentious. I generally advocate for giving people nearly all liberties, but I had no idea how to delicately answer this question, and neither did any of my staff. An advisor said we should check out what the US Constitution's Second Amendment (the right to bear arms) said about 3D-printers. We laughed, thinking it ridiculous to try governing a country with a 226-year old document in the transhumanist age.

People ask me all the time—since they know I'm not going to win the presidency (third party candidates never win)—if I'm enjoying the campaign. I've never thought about it like that.

I've only focused on one thing through it all—the same thing I've focused on with all my work for much of the last decade: I don't want to die. Like most transhumanists, it's not that I'm afraid of death, but I emphatically believe being alive is a miracle. Out of two billion planets that might have life in the universe, human beings managed to evolve, survive, and thrive on Planet Earth—enough so the species will probably reach the singularity in a half century's time and literally become superhuman.

The whole experience of life is crazy and beautiful and precious. We need to protect and preserve that life, at all costs. The only rational way one can do that is with technology and science, which is exactly the principle the Transhumanist Party was formed upon. Transhumanists are a people defined specifically by their love of life.

I've been lucky and grateful that so many people agree with me. Renowned gerontologist Aubrey de Grey has recently signed on as my Anti-aging Advisor. Millennial entrepreneur Riva-Melissa Tez is my Strategy Advisor. Jose Cordeiro, PhD, faculty at Singularity University, is my Technology Advisor. Former Democratic Congressional candidate Gabriel Rothblatt, son of transgendered billionaire entrepreneur Martine Rothblatt is my political advisor.

Even my wife, Dr. Lisa Memmel, a women's rights advocate and an ObGyn at Planned Parenthood, is a big supporter and member of my campaign. Together we are hoping to change the world and usher in an age where science, technology, and the right to do with your body what you want are not at odds with American culture.

Unfortunately, on many counts, transhumanism is at odds with our national culture. With an American population that is approximately 75% Christian and a US congress almost 100% religious, accomplishing my goal is no easy feat. Most Americans just don't care about the goals of transhumanism. Many subscribe to what I call a "deathist" culture, where they insist we must follow the rules of the Bible, die, and go to heaven to meet Jesus.

As an outspoken atheist (and apparently America's first visible atheist presidential candidate), none of this makes any sense to me. Despite this, I still insist my campaign and the Transhumanist Party welcome all religions and try to be respectful and open-minded of people's beliefs. After all, transhumanism is possibly the least discriminatory philosophy out there—it accepts anyone who wants to be a part of it.

None of this has calmed some of my detractors, though. Some people are downright livid about my ideas and campaign. I've even heard some Christian theologians suggest I might be the antichrist. As ludicrous as this talk is to me and others, it has prompted me to recently buy a bulletproof vest for public speeches. My share of hate email and death threats on Twitter are constant enough to warrant such measures.

These things dampen my spirit in multiple ways, since one of the central goals of the Transhumanist Party and my campaign is to advocate for taking money away from wars, violent activities, and defense, and instead put those resources into medicine to improve health, prosperity, and happiness for the citizenry. Violence, division, discrimination, and deliberately causing conflict are simply not on the transhumanist agenda whatsoever.

Transhumanism is a social movement, like environmentalism, that aims to unite people in a single direction and approach—an approach that is backed up by recent various reports showing that the world is getting better for all people. Over the last 30 years, science and technology have helped reduce infant mortality rates, given everyone longer lives, produced more jobs, lessened wars, improved general health, and made the world a better place. It makes sense then, to speed up the process of technological progress and embrace the transhumanist age, given how it improves lives.

Of course, when looked at it that way, not all Americans are philosophically opposed to transhumanism. And I think with the right diplomacy and gentle nudging, many more people in the states would come on board to embrace transhumanism. With that in mind, I'm on the lookout for new supporters. Aside from science and technology enthusiasts—those who make up the foundation of the Transhumanist Party—I am trying to acquire a new base of supporters that may allow transhumanism to break into mainstream politics.

My 2016 campaign strategy is to target three specific groups: atheists, LGBT people, and the disabled community. Collectively, they number about around 30 million Americans, and some of them are already present in large numbers in the transhumanist community and share similar values. I want to reach supporters of science and technology, and the main philosophical premise of morphological freedom—that you have the right to do with your body whatever you want so long as it doesn't hurt someone else.

Unlike the other Democratic and Republican presidential hopefuls, my campaign is not very technically political. Sure, I try to address questions on taxes, social security, international relations, and other typical candidate topics when asked, but I'm not trying to be spew political ideologies or subscribe to a political party. I don't care about leaning left or right, nor does most of the Transhumanist Party. We're here to offer the kind of change that affects society's entire existence and our rapidly evolving future. We want to convince government and people that a transhumanist-inspired country will not only benefit all, but be an exciting step for the human race.

For example, transhumanists want to reignite the space industry and send citizens all over solar system. We want to build massive seasteading projects where all flavors of people and scientific experimentation can abound. We want to create an artificial superintelligence that can teach us to fix all the environmental problems humans have caused. We want to declare war on cancer, Alzheimer's, and aging—not on drugs, disenfranchised minorities, or small oil-dependent nations.

We want to close economic inequality by establishing a universal basic income and also make education free to everyone at all levels, including college and preschool. We want to reimagine the American Dream, one where robots take our jobs, but we live a life of leisure, exploration, and anything we want on the back of the fruits of 21st Century progress.

My presidential campaign is a strange, tumultuous endeavor to undertake, knowing I have almost zero chance of winning the 2016 election. But I have my sights on other important tasks: growing the Transhumanist Party, developing policies that unite the nation under one techno-optimistic vision, and getting everyday people to desire unlimited lifespans via science. These are important jobs, and this is where my heart and mind are focused daily as I sojourn on the bumpy campaign trail.

****

25) If You Care About the Earth, Vote for the Least Religious Presidential Candidate

Earth Day came and went last week. And like years before, promises were made by governments and politicians to be better stewards of our planet. Just about any sane person realizes global warming is real and the damage humans have done to Planet Earth is substantial.

Most people believe a major step in the right direction to heal Earth's environmental crisis is to reduce humanity's carbon footprint and be more green—something being addressed in the recently signed Paris Agreement. While I applaud the collaborative effort and good intentions of the treaty, it's inadequate and doomed to failure. It's like bringing a water gun to a war zone. Nothing short of a mass-extinction event for humans can stop and reverse the environmental damage done or occurring to the planet. Billions of people around the developing world want the standard of life we have in America, and they're not going to stop for anything until they achieve that.

I don't know if the major US presidential candidates—like Donald Trump, Ted Cruz, or Hillary Clinton—are aware of this conundrum. And even if they were, the real question is: Can their politics, ethics, and religious beliefs handle it? Because sending out Christmas cards on recycled paper and giving tax incentives for electric cars is not going to pull us out of the toxic mess we've created on Earth. There's only one realistic hope to save the planet—and it comes from an unlikely place: technology. Radical technology. I'm talking CRISPR gene editing, transhumanism, and nanobots in every biological nook of the world. This will not be Kansas, anymore. And our current politicians will be freaked out by it.

The bright green future rests with disruptive tech. Consider this, for example: Twelve years ago, I used to work as a director at nonprofit wildlife organization WildiAid. In Cambodia, I went on undercover missions and helped bust and jail poachers who were causing wildlife—like tigers, Sun bears, and the Asian rhino—to go extinct. We did good work, but poaching is a nearly $20 billion business, and there's just no way a nonprofit organization (or even a dozen of them) could stop the demand for illegal wildlife, not when population growth in Asia is skyrocketing and poverty-stricken locals can sell a tiger for over $10,000.

But there are people who can save the endangered species on the planet. And they will soon dramatically change the nature of animal protection. Those people may have little to do with wildlife, but their genetics work holds the answer to stable animal population levels in the wild. In as little as five years, we may begin stocking endangered wildlife in places where poachers have hunted animals to extinction. We'll do this like we stock trout streams in America. Why spend resources in a losing battle to save endangered wildlife from being poached when you can spend the same amount to boost animal population levels ten-fold? Maye even 100-fold. This type of thinking is especially important in our oceans, which we've bloody well fished to near death.

As a US Presidential candidate who believes that all problems can be solved by science, I believe the best way to fix all of our environmental dilemmas is via technological innovation—not attempting to reverse our carbon footprint, recycle more, or go green.

As noted earlier, the obvious reason going green doesn't work—even though I still think it's a good disciplinary policy for humans—is the sheer impossibility of getting the developed world to stop… well, developing. You simply cannot tell an upcoming Chinese family not to drive cars. And you can't tell a burgeoning Indian city to only use renewable resources when it's cheaper to use fossil fuels. You also can't tell indigenous Brazilian parents to stop poaching when their children are hungry. These people will not listen. They want what they want, and are willing to partially destroy the planet to get it— especially when they know the developed world already possesses it.

So while I support green policies, there's no way such well-wishes will stop the future environmental degradation—nor will it reduce what already occurred. Even with a bunch of laws passed, or a massive cultural change, or everyone joining hands and singing "Kumbaya," we're in for dirty, toxic ride with the planet.

But don't lost hope. What can happen—and will likely happen barring a collapse of society—is our thriving modern world will devise transhumanist technological fixes to the issues at hand.

Take meatless meat for example—a powerful disruptive idea already here. About a third of the arable land on earth is dedicated to grazing animals. Much of that land was slash-and-burned to make way for cattle and other livestock. If we switched to meatless meat, which is made in a laboratory and tastes quite similar to real meat, we could stop the clear cutting and whole scale destruction of those ecosystems (not to mention we can avoid cruelly slaughtering 150 million animals a day for food). Additionally, if we can figure out ways to genetically grow back rainforests in weeks instead of years, we could stop creating all that greenhouse crap that goes into our air and atmosphere. With new genetic editing techniques, it's quite possible we could learn to speed up biological processes, like tree and plant growth.

It gets even better, though. Gene-editing technology like CRISPR/Cas9 could in the future make us cancer-proof, so even if we did have a depleted ozone layer from rainforest destruction and pollution, we'd never get cancer. Another alternative is to use CRISPR tech to give us sunburn-proof skin.

Of course, some prominent futurists like Ray Kurzweil and Peter Diamandis are predicting a word of nanobots and nanotechnology in 20 years or less, which could physically change the structure of our material existence. We'd simply create billions of mini-robots that inhibit everything and help recreate and protect natural beauty. The nanobots and nanotech will be able to eradicate pollution and serve green purposes on a nearly molecular scale.

Whichever way you twist it, radical science is the method that can help and save us most efficiently. We can use it to fix just about every stupid thing we've ever done to our planet and to its life forms. And best of all, science often doesn't just tackle problems by kicking the proverbial can down the road, but rather by eradicating issues completely—like we did with diseases like smallpox, or by inventing freezers so our food doesn't spoil, or by flashlights so we can see outside on a rainy night.

But here is our impasse. Not every government leader is inclined to use science to fix the environment—or even to help human being's health and longevity. Some politicians believe first and foremost in following their moral and religious ideology—like former President George W. Bush, who severely limited federal stem cell funding for seven years while in office because of his Christian values.

With the radical new age of CRISPR genetic tech upon us, I'm worried that conservatives like Cruz will also try to stop new technologies that will change our battle in combating a degrading Earth. It's likely Republicans will be against radical pro-green genetics or embracing environmentally-friendly cyborgism that seemingly counters their biblical view of the world. In this case, the Democrats—who are often far less religious than Republicans—may be more inclined to understand the necessity for changing our DNA or embracing nanotech to be become better climate-adjusters. In fact, maybe the Libertarians—who passionately insist on separation of church and state in public affairs—will be the most accepting of future tech to save the planet. Their nominee will be on all 50 states this election cycle.

Like it or not, genetic editing, nanotech, and cyborgism are a fundamental part of the future for how we deal with basically everything, and it will allow us to rewrite the coding—and hence alter the form and purpose—of the entire biological world.

Sadly, there are already calls for moratoriums being voiced over some of these types of science. If there's a ban on the research, how then will we learn to re-grow trees in weeks instead of years to replenish our fragile rainforests? How will we help build up endangered wildlife populations if the technology is outlawed? How will we become cancer-proof to higher UV rays if we can't experiment?

Whatever happens in the 2016 US elections, if you care about the environment—if you care about really making a difference to return this planet back to a pristine and green state—then vote for the politician who doesn't make their science decisions based on archaic religion and 5,000-year-old holy texts, but on what works and what is in the best interest of the people. The next person in the White House—especially if they manage to stay eight years—is going to make or break the issue of environmentalism—and the greatest hope for them is to stand strong with using radical science and technology as the their main weapon of change.

****

26) What It's Like to Counter-protest Christians as an Atheist Demonstrator at Both Major Political Conventions

Of the thousands of protesters at the 2016 Republican and Democratic conventions, one of the most noticeable groups were the born-again Christians. But this year—perhaps for the first time at national conventions—they were met with resistance from organized atheist and transhumanist protesters clashing against them.

With approximately a dozen atheists and transhumanists, my group of supporters engaged the Christians and campaigned in front of them. Sometimes wearing Transhumanist Party t-shirts and holding posters, we argued with them, blocked press from getting good pictures of them, and generally promoted to the public the power of reason over faith.

Other protesters, often fans of Bernie Sanders, joined our cause and helped protest against them. One of my favorites was an LGBT supporter holding up a sign with a penis drawing on it and the words "I love sinning" emblazoned just below it.

At the Republican National Convention, my team and I were able to temporarily block the born-again Christian procession attempting to make their way down the busy E 4th St.—right in front of the MSNBC live filming booth. Additionally, it caught the attention of GOP delegates and press as they were making their way into the nearby convention center.

You have to give the Christian protesters credit. They know how to make a spectacle and get their message out. You can hear them shouting scripture with their megaphones from 100 yards away. And their colorful six foot tall banners on poles are provocative and attention-getting. Some of the banners read "Homo Sex Is Sin" and "Stop Being a Sinner" and "Obey Jesus." Others banners quoted biblical scripture. Some talked of hell fire and the end of the world.

My team and I have some work to do to create better banners and be more organized, but it was a powerful start for many atheists like myself that want to counter religion's negative influence. I believe in the right for Christians to speak their minds, but I don't think nonreligious people should stand around quietly and just shake our heads at irrationality. Confronting religious deception is vital.

"It's important for atheists to make a stand against religious forces if atheists want a more secular world," says Mario Gruber, a writer and volunteer for the Richard Dawkins Foundation for Science and Reason.

We live in a nation where all 535 members of Congress, all eight Supreme Court justices, and our President are publicly religious and believe in an afterlife. For the life of me, I just can't see how that is authentic separation of church and state.

Politicians follow their religious inclinations when making policy decisions. A prime example of this comes from former President George W. Bush, who restricted federal funding for stem cell research during a majority of his presidency due to his religious convictions. America and the entire global medical field suffered dramatically because of that decision.

Hundreds of thousands have had worse health and some surely died as a result. Stem cell innovation continues to be one of the most important new treatments for a variety of ailments. Bush's restrictions have pushed research seven years behind where it could've been.

Many atheists and transhumanists question whether we should even bother protesting against fundamental Christians. Some have even asked me if it's counter-productive to declare myself an "atheist presidential candidate" to the media. However, I worry that passivity in the atheist community will ultimately keep us living in a Christian nation indefinitely. I would like to see America quickly grow out of its religiosity. I support people that embrace open-minded spirituality that isn't tied to any fundamental religious text or determines its rules bases on so-called historical prophets. Spirituality, as author Sam Harris puts it, is natural and helpful for a happy life:

"While spiritual experience is clearly a natural propensity of the human mind, we need not believe anything in insufficient evidence to actualize it."

The good news is that recent reports show the nonreligious growing quickly in America, especially amongst youth. I hope we continue down that path—and I believe we will. One sure way to do that is to not let the Christians be the only ones getting attention for spreading their message. Both national conventions had many thousands of journalists attending them, and it's important for atheists to show the media that our ideas are catching on and being actively represented to the largely religious public.

In the future, we'll have new ideas to counter protesting Christians, too. My wife is a medical doctor and works at Planned Parenthood, and as a transhumanist, I'm working on organizing a group of six foot tall robots to protest the religious demonstrations at abortion clinics. Additionally, The Transhumanist Party, which owns drones, is also looking into ways to use flying devices to carry atheist and anti-death banners in the sky, as well as loudly playing speeches and quotes from luminaries like Christopher Hitchens.

I know secular hackers who are working on creating atheist bots that spread secularist themes perpetually online. A whole new range of technologies will help the upcoming generation of atheist and transhumanist protesters to get their message out and make America more secular.

The future for atheists, transhumanists, and those who want to counter religion is looking strong. It begins with activism and standing firm against the faith-minded who wish to control us with archaic religiosity and unreasonableness.

****

27) The Future of the LGBT Movement May Involve Transhumanism

The other night my wife and I were reading to our 4-year-old daughter a children's book that we borrowed from the public library. We came to a section where two characters — both who were the same sex — began having romantic feelings for each other. My wife and I smiled — we have many good LGBT friends.

Later that evening after putting my daughter to bed, I began wondering about the future of the LGBT movement, especially after Tim Cook, Apple's CEO and probably the world's most influential technologist, recently said he was proud to be gay. It's certainly interesting to speculate on how sexuality, sexual orientation, and society's interpretation of it all will change over the next 25 years as we charge headlong into the transhumanist age.

It shouldn't come as a surprise to anyone that the LGBT movement and transhumanism have a lot in common. Nearly all transhumanists support the LGBT cause. After all, a desire to be free to alter, express, and control one's sexual preference and identity sounds like a transhumanist concept. Advocates of transhumanism aim to alter, express, and control their bodies and preferences too, except they emphasize doing it with science and technology. If you look closely, the two movements — especially some of their major philosophies — are practically different sides of the same coin, and each is poised to gain strength from one another in the future as radical technologies transform the species.

In the next 25 years, the human being will undergo a larger transformation of its evolutionary body than it has undergone in the last 100,000 years. Artificial hearts will likely become better than real hearts. Telepathy via brain implants will become an important form of communication. Men will be able to give birth with implanted uteri. Each of these technologies already exists in some form and will soon be more widely available.

The million dollar question regarding these technologies is whether we will be allowed to freely use them. After all, the United States Congress is basically made up of all religious politicians, some whose faiths derive from texts that forbid anything like LGBT practices or transhumanism. Transhumanist's main goals are to overcome mortality and become as free and powerful as possible using technology—in essence, to become godlike.

For ages now, society has largely been afraid of transformation, especially when it concerns the human body or sexuality. Even today, a dozen U.S. states still have anti-sodomy laws, and LGBT people are often killed in places around the world — sometimes stoned to death — for their actions and beliefs. While victories have been won in the 21st century, such as in California and other states where people of the same sex can now officially marry, massive inequalities and bigotry still exist.

In the future, transhumanist technology and science will compliment the LGBT movement and help push it forward in the face of continued social oppression and closed-mindedness. This is important, since LGBT people are devoted to freedom. They want to be free to do anything they please without condemnation so long as it doesn't hurt others. Transhumanists — a notable number who are LGBT themselves — want the same exact thing. And they can work together to better achieve their goals.

With the onslaught of new tech and advanced medical and surgical techniques hitting the market, it's likely the LGBT movement will involve more transhumanist issues in the future. For those who are conservative and resist change, this may prove challenging. Take cybersex and virtual reality, for example, where Facebook's Oculus Rift and haptic suits will allow people from all corners of the world to have group sex if they want. Or what about fembots and sexbots, which already represent a growing 100 million dollar market? In 10 years, some robots may be as sophisticated as humans. Do we give them rights? Can we marry them? What if they're gay? What if we program them to not know if they're gay or not?

"The world is shifting under our feet," says B.J. Murphy, a pansexual transhumanist, writer, and futurist. "In 15 years, conservatives and anti-gay people will look back at the LGBT movement and yearn for an adversary so simple in its demands."

B.J. Murphy is right. The future will be anything but simple. Already, within two decade's time, parents may choose to have designer babies without certain sexual organs. Is a uterus necessary if you have ectogenesis (use of artificial wombs)? Or does it just present extra cancer risk and, for some, decades of painful, crampy menstrual cycles? Alternatively, will some religions encourage some males to be born with genetically lowered sex drives so they may have a better chance at becoming celibate priests, a shrinking vocation in the U.S.? Finally, will some seemingly narcissistic people procreate only through cloning techniques? The bizarre questions of the transhumanist age seem endless — and they are already being asked by a growing number of people.

Frankly, I could see many humans in the future stopping physical sex altogether as cranial implant technology finds precisely the right means to stimulate erogenous zones in the brain — something researchers are already working on. Real sex will probably not be able to match direct and scientifically targeted stimulation of our minds. Such actions may lead to a society where male and female traits disappear as pleasure becomes "on-demand," and gene therapy is able to combine the most functional parts of both genders into one entity. Not surprisingly, some institutions like marriage may end up going the way of the dinosaurs.

The LGBT movement has found firm footing in the 21st century — a testament to the courage of its supporters. I applaud them and support their courageous efforts. As a transhumanist, atheist, and a politician, I stand ready to defend their freedoms and push their agenda forward, all the while knowing that the future will bring its own set of new challenges that none of us can easily foresee. In fact, the clash of civil rights in the transhumanist era may just be starting in a whole new way. Personhood, sexual freedom (virtual or not), and gender identity (or non-identity) will soon take on unprecedented roles in society, spurred by radical innovation and changing stereotypes of what it means to be a human being. For me, the wildcard of the future is not in society, but in the transformative technology that we invent and embrace.

****

28) We Must Cut the Military and Transition into a Science-Industrial Complex

Many Americans subscribe to the annoying belief that our nation's military-industrial complex is the surest way to remain the wealthiest and leading superpower in the world. After all, it's worked for the last century, pro-military supporters love to point out.

However, America's dependence on warmongering may soon become a liability that is impossible to maintain. Transhumanism, globalization, and outright replacement of human soldiers with robots are redefining the county's military requirements, and they may eventually render defense budgets far smaller than those now. To compensate and keep America spending approximately 20 percent of the federal budget on defense (as we have for most of the last few years), we'll either have to manufacture wars to use all our newly-made bombs, or find another way to keep the American economy afloat.

It just so happens that there is another way—a method that would satisfy liberals and conservatives alike. Instead of always spending more on our military, we could transition our nation and its economy into a scientific-industrial complex.

There's compelling reason to do this beyond what meets the eye. Transhumanist technology is starting to radically change human life. Many experts expect to be able to stop aging and conquer death for human beings in the next 25 years. Others, like myself, see humans merging with machines and replacing our every organ with bionic ones.

Such a new transhuman society will require many trillions of dollars to satisfy humans ever-growing desire for physical perfection (machine or biological) in the transhumanist age. We could keep our economy humming along for decades because of it.

Whatever happens, something is going to have to give in the future regarding military profiteering. Part of this is because in the past, the military-industrial complex operated off always keeping a few million US military members ready on a moment's notice to travel around the world and fight. But there's almost no scenario where we would need that kind of human-power (and infrastructure to support it) anymore.

Increasingly, small teams of special operation soldiers and uber-high tech are the way America fights its wars. We just don't need massive military bases anymore, nor the thousands of companies to support the constant maintenance of ground troops. Such a reality changes the economics of the military dramatically, and will eventually leave it a fraction of its size in terms of personnel and real estate.

We'll still have the need for technology to fight the wars and conflicts we entangle ourselves in, but it'll be mostly engineers, programmers, and technicians who wear the uniform. The coming military age of automated drones, robot tanks, cyberwarfare, and artificial intelligence just doesn't require that many people. In fact, expect the military not just to shrink, but to mostly disappear into ones and zeroes.

Many people think that the beast of a military-industrial complex—made famous by President Dwight Eisenhower's warning against it in his farewell address—appeared only in the last 50 years. However, others persuasively argue that America has been at war 93 percent of the time since the US Declaration of Independence was signed in 1776—so it's been with us from the beginning.

In liberal California where I live, such facts annoy just about everyone I know—except, of course, those who are shareholders and beneficiaries of the defense industry. Thankfully, despite Congress being led by mostly older white religious men, the younger generation clamors for an improved America—one that can keep its economies running smoothly in a more peaceful way.

This is where the scientific-industrial complex comes in and could satisfy most everyone. And best of all, a society of science requires actual people. Lots of them: nurses, scientists, start-up CEOs, designers, technologists, and even lawyers. The advent of modern medicine to treat virtually every ailment—and the whole anti-aging movement, in general—affects all 318 million Americans. Over half of us suffer from health issues that can be improved but often aren't, for a variety of reasons. For example, the US Census Bureau reports that 40 percent of people over the age of 65 suffer from a disability—and for two thirds of them, it's mobility-related issues. And millions are already racking up the symptoms of heart disease that will kill them. And a younger generation is just waiting to explore bionics, chip implants, and how to upgrade their genes to avoid health problems in the future. All this means we have the fodder to reshape the American economy from a militaristic-based one to a type that thrives off scientific and medical innovation.

Instead of spending American money on sending our soldiers to risk their lives for the whims of war, we could be giving civilians the medicine and healthcare they need to live far better and longer. And living longer has unseen benefits, too. In the future, bonafide transhumans won't have to retire if they don't want to. Their bodies will be ageless and made so strong through technology that work and careers may continue indefinitely—and therefore, so will paying taxes. Transhuman existence is a self-fulfilling economic-boom prophesy for both individual and country.

To help create this new mindset in society, I recently delivered a Transhumanist Bill of Rights to the US Capitol as part of my presidential campaign tour. Article 1 of the bill, among other things, aims to establish that a nation would provide a universal right via science and technology for citizens to live indefinitely if they wanted. This, of course, takes socialized medicine one step further, and doesn't just mandate that the government is interested in your well being, but that it's ultimately interested in your permanent survival.

If a nation was to embrace such a universal right to live indefinitely, it would forever change how a nation looks at the individual lives of its citizens. What would follow is a nation's intense build-out of how to improve the health, longevity, and well being of its people. Additionally, the institutions that are constantly drawing on America, like social security and welfare due to disability, would be less taxing.

Currently, the US Constitution (which I personally think needs a significant rewrite for the 21st century) is overly concerned with protection of national sovereignty—which is one major reason why the military-industrial complex is allowed to grow undeterred. If the US Constitution was endowed with precise wording to also protect an individual's health, well-being, and longevity, then a scientific-industrial complex could rise. This new monster institution would legally be mandated to provide the most modern medicine, technology, and science possible to its people.

Shamefully, the Iraq War will cost the US $6 trillion dollars by the time we're actually done paying all our bills—despite the fact that it's highly questionable whether Iraq was ever even a serious national security issue. However, our country undeniably faces a serious national security issue today—in fact, I'd call it a full blown crisis. Nearly 7,000 Americans will die in the next 24 hours from cancer, heart disease, diabetes, aging, and other issues. And the same amount of people will die tomorrow and the day after.

Overcoming disease and aging in the transhuman age will inevitably occur. The question is not if, but when? The answer lies in how much our nation is willing to spend on scientific and medical research—and how soon. But so long as it continues to spend money on the military instead of citizen's health, human beings will die—which is ironic since it's the military that is supposed to protect us (and not inadvertently sabotage us by swallowing funding for bombs instead of medicine). All we need do as a country is change the direction of our spending, from defense to science. If we can transform America into a scientific-industrial complex, we'll still be able to keep our economy chugging along. Let America's new wars be fought against cancer, diabetes, Alzheimer's, and aging itself. It's a win-win, except for body bag and casket makers

29) Transhumanist Rights are the Civil Rights of the 21st Century

Maitreya One, a black futurist and hip-hop artist living in Harlem, steps off the Greyhound bus on a warm morning in Montgomery, Alabama. Wearing sunglasses and a backwards-facing baseball hat, he eyes the film crew covering his arrival. I walk up to him and give him a hug. I'm excited he's here.

Maitreya is a civil rights link from the past to the future—and one of the few African-American transhumanists I know. He is stepping off one bus in Montgomery—whose roots are tied to the spectacular Freedom Riders who challenged segregation laws in the early 1960s—and onto another: the coffin-shaped Immortality Bus, whose mission is to spread radical science and promote life extension and transhumanist rights.

Like others in the burgeoning transhumanism movement, Maitreya supports becoming a cyborg in the future, and he knows the coming controversy over such aims may end up as challenging as the civil rights era battles over racism.

To transhumanists—some who want to become new biological species and others who want to become machines—a new civil rights age is looming. We can already see the start of it with numerous calls for a research moratorium on human genome editing—a scientific feat that took place in China in 2015.

But there are many more difficult questions beyond directly modifying the biology of the human being. Should humans be able to marry robots? Should sophisticated artificial intelligences be given personhood? And are crimes committed in virtual reality punishable by jail time? The questions are endless.

Navigating the future of transhumanism is indeed thorny. And even though a lot has changed and improved in the 55 years since black and white Freedom Riders risked their lives— arriving in Alabama in buses to challenge Jim Crow segregation practices—bigotry, traditionalism, and closed-mindedness is alive and well. And this conservatism may hold us back.

"We need morphological freedom—the right to do with your body whatever you want, so long as it doesn't hurt anyone," Maitreya says. "We need real policies of justice that serve everyone and do not discriminate against new ideas."

On the Immortality Bus, we head to downtown Montgomery to the Freedom Riders Greyhound station, which is now a museum. Outside are photos plastered to the building where white Southerners (some who belonged to the Ku Klux Klan) once attacked bus riders who wouldn't segregate on buses or in bus terminals.

I ask if Maitreya thinks the future of transhumanist civil rights might become as violent as this.

"I hope not," he answers. "I hope this was just a bad period of history in America. I can't even believe all this happened because white and black people wanted to sit next to one another. It's ridiculous."

I agree with him, but I'm skeptical whether future civil rights won't also have its share of violence as the world progresses forward. While my travels in the South as a pro-technology and non-religious U.S. presidential candidate have been met with kindness and curiosity, it's also been easy to quickly turn off people.

As soon as I tell people I have a chip implant in my hand, opinions of my campaign seem to quickly change. Religious people dislike any type of technology that brings ups questions of Revelations in the Bible or the Mark of the Beast. Implants, a classic transhumanist technology, seems to provoke just those exact ideas.

Unfortunately, everything transhumanists are trying to accomplish—from conquering death with science, to merging with machines, to becoming as powerful as possible via technology—conflict somewhat with biblical scripture and conservatism. The word transhuman means "beyond human" and that's what most transhumanists are striving towards. Naturally, that is going to rub the wrong way on many people who believe in the sanctity of the natural human body and traditional human experience.

Like any new potentially society-changing movement, transhumanism has its work cut out for it with future civil rights. The concept of personhood used to be a simple one, but with artificial intelligence and robots that can already nanny our children and cook dinners, we will soon see a time when courts must decide how far to take these ideas.

For example, if an intelligent robot makes money, should it be taxed? Or will all robots fall under a nonprofit entity status? Robots like this will likely arrive in households and the workforce before 2025, so we're not talking some distance future, but something only years away.

In the biology realm, the issues are already here. Cloning is banned in many states. And stem cells derived from fetuses— a classic transhumanist pursuit—is frowned upon by many. Cryonics is illegal in some places in the world. The LGBT movement—which many transhumanists strongly support—will also be affected as gender reassignment surgery becomes an easy procedure. Many transhumanists, including myself, believe we'll eventually arrive to a genderless world, made possible via science. Perhaps even more controversial, artificial wombs will challenge maternity, and also upturn abortion clashes. All this, besides the fact that men will soon be able to have babies themselves with uterus transplant surgery.

The coming conflict of advancing technology vs. human rights is a massive one. Already, we've seen some initial banning of Google Glass in public places, which I suspect made Google not push hard for the success of that product. Additionally, the U.S. National Institute of Health (NIH) recently reaffirmed its ban on gene editing of embryos. And laws of virtual persons are being discussed and applied to Second Life and similar places.

Easily, the biggest transhumanist issue in the future is robots taking jobs. Even South Korea has already replaced some prison guards with robotic guards. And some hotels now use robots, as well as Lowes, the home improvement giant. All these issues fall under the umbrella of transhumanist civil rights—and nothing in the young field is simple or for certain.

Some anti-tech naysayers and luddites are screaming to slow down technology before it gets out of control. Others, like myself, believe technology will only help the planet. History shows—at least in the last 30 years, according to the World Bank—that science and technology have been making lives longer and better for virtually everyone on the planet.

Despite my optimism, I still understand the need to tread carefully. We are in new territory, and endless amounts of discussion must reach the highest levels to find the best path. Unfortunately, even during the 2016 presidential cycle, virtually no politicians are discussing some of the most pertinent issues at hand, like designer babies, or A.I. controlling nuclear arms, or whether 3D printing of guns and bombs is legal and should be encouraged.

As a transhumanist U.S. presidential candidate, these issues are my main focus. In fact, the main goal of the Immortality Bus is to spread a newly written Transhumanist Bill of Rights that covers many of these issues. But as a third party candidate with virtually no chance of winning, much of it falls on closed ears.

It's my hope, though, that through the music of people like Maitreya One or other initiatives of transhumanists, that more and more people will start to discuss the future—before it arrives and slams into us. That way, we might be able to avoid confrontation by understanding the possibilities before the next new civil rights battles emerge.

****

30) Is it Time for Fast Track Atheist Security Checks at Airports?

My last month has been increasingly busy as my US Presidential campaign picks up speed, and I participate in more and more events. In the past few weeks, I've made speeches in London, Vancouver, and Palm Springs. I live in San Francisco, so each of these destinations required air travel. Unfortunately, the travel also meant I had to endure long-lined security checks to board airplanes.

We all know the dreaded procedure. Line up like cattle. Then when near the X-ray screening conveyor belt, pull out your laptop and place it in its own plastic tub. Then take off your shoes (and in my case a belt). Then throw in your wallet or purse too, followed by keys, smartphones, and carry-ons. If you have really young kids like I do, then the drama sometimes involves close examination by a surgical-gloved TSA officer scoping out dubious baby formula in a bottle. In fact, in London's Heathrow Airport, they even asked me to take off my wedding ring—just in case it might belong to Suaron, the Dark Lord of Mordor.

In my six flights in the last month, I never managed to get through any security check in less than 40 minutes. Naturally, I wondered if it really had to be this way. Yet, when I looked around me in the security check lines, I found my answer.

A Sikh man in a turban was in front of me. In another line was a Muslim women wearing a black burja. Behind me a Catholic priest in a robe carried a worn leather satchel. Sure, religiosity makes some people fundamentalists. And, historically, such beliefs have been used to perpetrate great harm, including the bombing or highjacking of airplanes.

As an atheist, my beliefs do not make me a fundamentalist—in fact, they make me exactly the opposite. If I believe in anything sacred, it's the scientific method, the rational way of discovery which insists one is able to improve upon some set of facts by systematic observation and experiment, but that improvement is probably still way off from any ultimate truth. Atheists are comfortable without knowing all the answers, and that is something that deeply defines us.

I couldn't help but wonder if I really had to wait in hour-plus lines since no atheist—so far as I could discover—has ever been accused of bombing or highjacking a plane for his lack of believing in God. Atheists aren't attracted to terrorism since they're too level-headed to believe they know all the answers to the universe. They don't need to defend or promote strict ideologies, especially archaic ones.

However, numerous aviation terrorist attacks have been committed in the name of religion—many during peacetime. And it's not just Muslims that are doing it, either. Christians have done it, as well.

As an aspiring politician, I strive to improve society by applying statistical analysis to decision making. I look at numbers and facts, and try to logically create policy that achieves the greater good for society, especially in a scientific transhuman way. I'm wondering if maybe it's time for an atheist security check line for the nearly three billion people that fly every year in the thousands of airports worldwide. After all, nearly 15 percent of the world's population is godless or nonreligious. Millions of productive hours (the equivalent of at least hundreds of millions of dollars) are being spent needlessly by atheists in security check lines every year—all because a number of religious people may use planes as terror weapons.

While this isn't and has never been an actual policy of mine, I envision a fast track line for atheists at all busy commercial airports, with only visual screening from a distance by TSA personnel. To use such a line, a traveler would simply have to publicly check that they're an atheist when getting ticketed, and then off they'd go through security with no wait.

Such a no-wait line would be interesting for a number of reasons. Firstly, religion is declining in America, anyway. The recent Pew Research Center study shows notable drops in formal religious beliefs. So ultimately, such a fast track line would force wishy-washy believers, or those who are religiously apathetic, to publicly write off their faiths and God in the name of not wanting to wait in exhausting security lines.

On the economic front, less TSA people would be needed, giving tax payers more money in their pockets.

One of the main reasons I'm running for US President is to try to get this country to be more rational—in this case: considering whether there's a necessity for airport security checks for atheists. Since religion and terrorism are statistically connected, perhaps we should leave the two to themselves to work out their qualms. But for atheists like myself, to wait in dreaded airport security check lines for about five hours in the last month, is unacceptable. There must be better methods to move secular society forward, and it could start with a conversation about fast track atheist lines that force people to think about religion's true social and economic cost.

****

CHAPTER VI: SECULAR VISITS & EVENTS

31) The World's First Atheist Orphanage Has Launched a Crowdfunding Campaign

BiZoHa Orphanage, which bills itself as "the world's first atheist orphanage," has launched a crowdfunding campaign. Its aim is to build a deliberately godless safe haven for homeless children in Kasese, Uganda. The proposed orphanage will offer refuge to children, most of whom have lost their parents to AIDS, and will become part of the broader science-based Kasese Humanist Primary School (KHPS).

The secular-themed KHPS is already aligned with Foundation Beyond Belief and Atheist Alliance International, and has garnered acclaim from renowned atheists. Its new atheist orphanage will be located near the Rwenzori Mountains and the Congo border. The crowdsourcing campaign is seeking start-up funds of approximately $4,500—that amount will fund the construction of a home for 15-20 orphans, provide it with furniture, and feed the orphans for one year.

"BiZoHa Orphanage intends to be economically self-sufficient within one year of opening its doors," the project's co-coordinator, Hank Pellissier, of the Brighter Brain's Institute, told me. "This goal will be achieved by selling corn, beans, cassava, peanuts, and lettuce grown on its 7-acre crop farm, which is part of the proposed orphanage."

The project began when Pellissier visited Uganda earlier this year and formed a friendship with KHPS director, Bwambale M Robert. Robert, an Ugandan, was orphaned when he was five years old. He worked his way through high school as a barber, and then went on to study Biology in college. He is the author of the forthcoming book, Orphans of Rwenzori - A Humanist Perspective.

"There's a very real need for orphanages in Uganda," Robert wrote me in an email. "We have 3.5 million orphans, 9 percent of the population. Children are parentless due to AIDS, civil war, violence, accidents, and abandonment. Up to 70 percent of orphans become criminals as adults. Among girls, 60 percent end up in prostitution, where the HIV/AIDS rate is 37 percent."

Education, Robert says, is limited. "When they 'age out' of orphanages, many become 'street kids' sniffing glue, stealing, scavenging in garbage dumps, and begging. Many are subjected to illness, filth, malnutrition, sexual abuse, and child trafficking—which is slavery."

Robert knows the experience firsthand. "Myself having grown up as an orphan since the age of five has made me experience all kinds of life and encounter many challenges," he told me. "Because I'm now better off than the orphans in my country, I feel it's an opportunity to join hands and rescue them so that they can live fulfilled lives like the rest of the people."

Orphans who attend BiZoHa will get a shot at education, and be able to attend the Kasese Humanist Primary School, which Robert founded in 2011. Many graduates of KHPS advance to secondary schools and even universities, like Robert did.

"That's why we need a declared atheist orphanage. We want to teach the kids there about science, secularism, and what it means to be an atheist. We also need to stand up for the values we believe in."

"There are a number of secular schools and orphanages around the world," atheist futurist B.J. Murphy told me. "But many atheists don't need to label themselves so directly as 'atheist.' The same is the case with naming of orphanages. We are not like the religious trying to convert people to not believe in God."

Murphy hits the heart of the issue. In the US, religious iconography is rampant—God is in our money, our politics, and, yes, our schools, even the public ones, where students pledge allegiance to a nation "under God."

"That's why we need a declared atheist orphanage," Pellissier says. "We want to teach the kids there about science, secularism, and what it means to be an atheist. We also need to stand up for the values we believe in."

I tend to agree with Pellissier, and I've gently argued before for an open public discussion on how much religious indoctrination should be allowed to be taught to young, developing minds. In a world where religious child soldiers carry AK-47s, Christian kids bully their gay peers, faith inspires infant genital mutilation, and teenage Jihadists become suicide bombers, I believe we must take more care to protect the young of the world and their sponge-like brains.

However, even as the Transhumanist Party's openly atheist 2016 US Presidential candidate, I still believe that if adults wish to pursue fundamental religious ideas, they should have the right to do so—so long as they don't harm others. I also think, even in an atheist orphanage, it's okay to share some religious ideas along side atheist perspectives, so long as there is a overriding emphasis on reason.

"The question of believing in God and religion should be a choice, and not something programmed into us," Murphy said.

Pellissier, who considers himself open to spirituality, says that orphans who are religious will be welcomed into the orphanage. But the education, culture, and emphasis will be on an atheist and secular experience.

Regardless of what kids end up believing when they become adults, the BiZoHa Orphanage will provide its street kids with an unprecedented opportunity. The orphanage's hope and goal—similar to Robert's experience—is that more opportunities will spring up for these orphans as they make their way through life with a sound education and upbringing that focuses on reason.

The atheist orphanage will carry the motto: "With Science, We Can Progress."

****

32) I Visited a Church that Wants to Conquer Death

Many people think of transhumanism — the belief that humans can evolve through science and tech — as a secular movement. For the most part it is, but there are a number of organizations that aim to combine science and spirituality together.

One of the largest is the Church of Perpetual Life, a brick and mortar worship center near Miami, Florida that looks like any other church. It has a minister, a congregation, and church activities. The only difference is this church wants to use science to conquer death.

I was asked to speak at a Church of Perpetual Life service while traveling across America on my Immortality Bus— a coffin-like campaign bus I'm using during my run for president of the US (under the guise of the Transhumanist Part). Services at the Church of Perpetual Life don't revolve around worshiping a deity. They're passionate exploration of life extension research. It's a group of people that want to live forever, but also want belong to a spiritual community.

Conversations are centered around how humanity can improve itself through science, how we can overcome death with technology, and how suffering can be broadly eliminated.

The church itself welcomes people of all religions, and sometimes explores concepts of a benign creator in very nonspecific terms. But mostly church services are dedicated to hosting invited speakers who make presentations on the current status of the anti-aging field. For example, gerontologist Dr. Aubrey de Grey, a Transhumanist Party anti-aging advisor, spoke there recently. So did entrepreneur Martine Rothblatt of Terasem.

The Church of Perpetual Life—whose symbol is a fiery phoenix—was originally founded by multi-millionaire Bill Faloon and his business partner Saul Kent. Faloon is known widely in the transhumanist community for being a cryonicist, and he has helped fund many life extension projects.

A cornerstone of the philosophy of the Church of Perpetual Life is its interest in the 19th century Russian prophet Nikolai Fyodorovish Fyororov, considered by some an early transhumanist. He believed we could follow science to become our best selves, and that is was the task of humanity to conquer death and unite humans in love and peace. The Church of Perpetual Life considers him a prophet.

Major church services take place about every month, and sometimes more frequently if a longevity speaker is in town. Neal VenDerRee, the certified minister of the church, presides over the sermons. He is also the main go-to person of the 500+ person congregation. VanDerRee and I spoke a number of times on Buddhist philosophy, which both of us appreciate greatly.

On the night I spoke, the 38-foot-long Immortality Bus was parked by the church entrance, with flood lights hovering over it. Because the bus resembles a giant coffin (to remind people we should all be working on overcoming death), the church decided to put a spotlight on it after the sermon for the 60 or so churchgoers.

Because I'm a US presidential candidate, my speeches are almost always political. But I promised VanDerRee I wouldn't speak at all about politics. So instead, I spoke on how important it is spread transhumanism to the general public. There were discussions about the implant I have in my hand, stem cell technology, and whether mind uploading is possible. I believe it is.

After my speech, Bill Faloon gave a short passionate talk on the dangers of high blood pressure, and transhumanist Maitreya One performed a short rap song about longevity. The evening ended with drinks and dinner, as well as visits aboard the Immortality Bus.

Leaving the Church of Perpetual Life made me think about my atheism. After being raised a Catholic, and even attending Catholic school where religious dogma was drilled into me, it was refreshing to be inside a church and feel part of a spiritual community without all the threats of damnation.

A church that asks nothing from you and hopes to end death for all humanity using science—now that's something I can support.

****

33) I Visited a Community Where People Upload Their Personalities to 'Mindfiles' so They can Live On After Death

As transhumanism and its quest to achieve indefinite lifespans through science moves more into the mainstream, questions of whether there's any room for spirituality in the movement abound. The answer seems to be a resounding "yes."

Transhumanism spirituality revolves around how technology can impact the greater truths our species faces, including whether a God exists or not, or even theistcideism—the idea that God might have once existed, but no longer does.

Terasem is one of the largest transhumanist communities in the world, exploring these questions while embracing radical science and technology to overcome death. Founded by multi-millionaire transgender tech entrepreneur Martine Rothblatt, Terasem is based in Central Florida, right on the Ocean. From the outside, it looks like a normal building, but inside it looks like an ashram—and most people, including employees, call it that.

As a transhumanist presidential candidate, my journey to Terasem began when I was invited to speak at Terasem's Annual Colloquium of the Law of Futuristic Persons in Second Life. While I knew about Second Life — the online environment where people build virtual worlds — embarrassingly, I'd never actually been in it.

Lori Rhodes, who helps manage Terasem, told me not to worry, and invited my campaign crew to visit.

We were joined by Terasem Pastor Gabriel Rothblatt, Martine's son and also a former Democratic candidate for Congress. Gabriel is a well-known spiritual transhumanist.

Gabriel told me, "The end goal of Terasem is similar to other religions — these ideas of joyful immortality in the afterlife. But for us it's not simply a spiritual concept, it's a mechanical challenge. Technology could one day make this a reality through digital backups - the idea of transferring a person's consciousness on to a hard drive, which could then be placed into quasi-utopian conditions. Heaven could be a virtual reality world hosted on a computer server somewhere."

One of the core tenets of Terasem is its belief in mindfiles, or digital compilations of people.

People in the Terasem community upload details of their daily lives and thoughts in hopes of recreating themselves one day, or just leave a lasting legacy in the future digital world. A few hundred people have mindfiles, and their information is kept securely on servers at Terasem, and backed up elsewhere around the world.

Anthony Cuthbertson, an embedded International Business Times who joined the Terasem tour, wrote:

"The so-called mindfiles include personality profiles, biographical information and memories in the form of photos and other media. For now it is more like a digital scrapbook but it is hoped that advances in artificial intelligence could one day turn these mindfiles into what can be considered human consciousness. Indeed, one of its mantras is 'software people are people too'."

The property includes a few large antennas, where the mindfiles are spacecasted, on the off chance that other life forms might pick them up (also just to get the data circulating throughout the universe, which in itself is a small piece of immortality for humans).

In addition to mindfiles, many Terasem supporters are, naturally, believers in other life extension technologies. Most Terasem members also want to use cryonics, where dead patients are frozen for long periods of time as they wait for new medical technologies to revive and cure them.

If the the cryonics freezing procedure accidentally damages parts of a patient's brains and memories, the mindfiles could theoretically be useful in helping determine who they are and once were.

With a little help from artificial intelligence, mindfiles may one day have a mind of their own. A hanful of companies, including Eternime and ETER9, are attempting to create mindfile-like platforms that can use artificial intelligence to post as you indefinitely on social networks. For some, this is scary stuff, but for others this might mean watching the avatar of one's deceased friend, parent, or child continue some form of existence, even if it's just in social media.

At Terasem, staffers prepare an amazing organic vegetarian lunch. A few filmmakers are present, recording everything. The lunch conversation is light, and I can't help but notice the large antennas just in view of the windows to the sea. It makes me wonder if the video being taken of our lunch — and even the words I've written here — will one day be used to help reconstitute someone at the table, or even myself. If so, I'm all for it.

****

34) I Visited One of the Largest Megachurches in America as an Atheist Transhumanist Presidential Candidate — Here's What Happened

As part of my 2016 US Presidential campaign representing the Transhumanist Party, I've spent much of the last month in America's highly religious South, traversing the Bible Belt and spreading the news that soon radical science and technology will overcome biological death.

Some experts are predicting a brave new world where gene editing, robotic hearts, and cranial implants may forever change humans into something transhuman.

No matter how you twist it, such concepts don't easily jibe with biblical scripture.

In my travels, I expected resistance to my message. Instead, people in the South have graciously offered curiosity and even support of my strange campaign. While I imagined we'd have rocks thrown at our bus, instead we got lots of people wanting selfies with us and local TV crews covering the tour.

My coffin-shaped bus recently made a visit to Alabama's Church of the Highlands. This nondenominational Christian megachurch, which has a dozen campuses and 32,000 members, is the largest church in the state.

Pastor Kyle Cantrell first encountered my crew and I while we were checking out the massive speaking pulpit where Sunday services are held.

Because I'm an atheist presidential candidate, my wife is a physician at Planned Parenthood, and I endorse microchipping humans for a variety of reasons (I have a chip implant myself), I was thankful my small traveling crew was treated so well.

Pastor Cantrell took us to the main chapel — which is separate from the Sunday service auditoriums — and I pressed him with questions about how technology might affect religion in the future.

One topic we discussed in detail was virtual reality. I asked him if he thought it might be used in teaching people about God. He told me he couldn't see any reason why virtual reality couldn't be used to facilitate Christian understanding and preaching. Cantrell seemed to think it might especially be useful for the physically disabled who might not otherwise be able to easily make it into the church.

One pertinent topic I've thought about before — especially since I was raised Catholic — is whether robots and artificial intelligences can be saved. I pointed out the Pope had recently mentioned that perhaps aliens could be saved if they existed.

"It's really the first time I've thought about whether robots or artificial intelligence could be saved," Cantrell told me, "but it's an interesting concept."

A time is coming, perhaps in as little as 20 years, when scientists will create intelligences as sophisticated as humans, and even I'm curious whether they will embrace spiritual values. I don't mind if they do — so long as they are not fanatical about it — but I do hope they will always hold reason and the Scientific Method as their highest codes.

I've been openly writing about my atheism for many years now, though if I really had to peg down my beliefs, I'd probably lean towards being a theistcideist— someone who believes a superintelligence like God may have existed at one point, but probably ended its own existence to give free will to the universe.

Most transhumanists embrace some spirituality, including myself. And despite my secularism, I'm quite certain that other intelligences are out there in the universe that are smarter than human beings. In an expanding universe that is almost 14 billion years old and may have 20 billion habitable planets, it's egotistical to think that humans are the only entities to evolve with advanced intelligence.

In the end, technology is changing the human race so rapidly that controversial topics like abortion, the existence of heaven, and saving the souls of robots may not matter in 30 years time. Technology may literally eliminate the questions.

For example, abortions may drastically decline due to artificial wombs, better forms of birth control that we control with our smartphones, and a possible overall decline in biological sex as virtual sex and sex chip implants become better than the real thing. And far fewer people will worry about whether heaven exists if science can conquer death and reverse aging.

Lastly, robots and AI will probably outperform human intelligences, likely teaching us about spirituality and possibly even about a superintelligence like God that may already exist — or did once exist.

****

35) I Visited a Facility Where Dead People are Frozen so They can be Revived Later

Over 100,000 people die each day globally. Why don't more of us consider cryonics — the practice of freezing the clinically dead in the hopes of bringing them back to life at a later date — as a way to avoid death?

As part of my 2016 US Presidential campaign representing the transhumanist party, I had a chance to stop in at Alcor Life Extension Foundation, the world's best-known cryonics facility, to find out.

You've probably seen cryonics before. Hollywood loves to use it in movies. Mike Myers (as Austin Powers), Woody Allen, and Mel Gibson are just some examples of people who have been "frozen" on the big screen.

Cryonics — and the field of life extension — has also been in the news a lot lately. A recent New York Times article featuring an Alcor patient generated discussion across the Internet.

Most cryonicists would not call frozen patients dead. They say patients are temporarily beyond the help of modern medicine, and that cryonics is the final attempt to provide emergency healthcare. Cryonics, they argue, is actually saving a patient by buying them time for science to catch up to the point where they can be revived.

"Death is not a moment, but a process, where an individual goes from a state of health, through many steps which end up becoming irreversible by modern means. It is not an absolute event. It is almost entirely dependent on the skills and means of the rescuer, and as we know, those skills and means improve over time," says Christine Gaspar, a RN and President of Cryonics Society of Canada and CEO Biostasis Canada.

I visited Alcor while traveling cross-country aboard my Immortality Bus (a campaign bus that resembles a coffin). Dr. Max More, a philosopher and the CEO of Alcor, gave me and my Transhumanist Party volunteers a private tour at the nonprofit's Scottsdale, Arizona facility.

One thing that struck me about Alcor was its size. It's not a small shop housing a few dead people in big steel tubes. It's a giant medical facility, complete with offices, surgical bays, laboratories, conference rooms, and of course, a large, highly secured hall for the cryonic tanks, known as dewars.

More oversees many of the cryonics procedures, and has a medical and scientific advisory board to look after operations. His team includes medical doctors, paramedics, and surgeons. 138 patients have been placed in cryonic suspension at Alcor so far.

Among these patients are baseball legend Ted Williams, transhumanism advocate FM-2030, and James H. Bedford, PhD., the first patient to be cryopreserved back in 1967.

"FM-2030 is a good friend of mine," More tells me.

That comment made me wonder about the immense challenges and commitment of being responsible — literally —for the existence of one's good friends. It seems overwhelming, but More, a fit 51-year-old, is up to the challenge. He is a steward for the transhumanist community — overseeing the bodies of friends and their families as they grow too old for science to help them. He acts as their guardian, advocate, and spokesman.

The process of cryonics begins with signing up for the service and gaining membership at Alcor or one of the other few cryonics facilities in existence. Ideally, a patient dies near a cryonics facility, so that they can be immediately cooled and prepared.

This is the ideal condition for diffusing cryoprotectants in the brain: Cooling the patient down to liquid nitrogen temperatures in the most controlled method possible, so that brain neurons containing memories and (hopefully) identity can be protected and preserved.

Many cryonicists wear "dog tags" or other identifying jewelry that show they require cryopreservation immediately after pronouncement of death — and medical professionals are supposed to respond to that. Dr. More told me one person even had instructions tattooed on himself so that they could be easily seen. Patients that aren't transported to a cryonics facility within a few hours of death are thought to not be preserved in ideal conditions.

Patients are placed in a bath of ice for transport and infused with chemicals to help preserve their cells and tissue structures in a process called vitrification. This, hopefully, eliminates the formation of ice crystals that can puncture cell walls and destroy the cells themselves. Later, either the head or whole body (depending on the preference of the patient) is transferred into a giant dewar filled with liquid nitrogen. Preserving just your head at Alcor is about half the price of the body, coming in around $80,000 plus minimal annual fees.

The science at Alcor and in cryonics is constantly improving, according to More. He tells me the new techniques they've been using the in the last 10 years are better at staving off the ice crystals that scientists suggest might lead to damage of the brain.

Some experts believe that patients could be revived in as little as 50 years, though there is no definitive way to prove this.

When I first began my Immortality Bus tour, I considered transforming the bus into a cryonics dewar to raise attention to life extension issues. But so few people knew about cryonics that the bus's effect on the public would be muted. So I chose to make my bus look like a coffin, and most people get it right away.

Because I'm in excellent health, and cryonics is expensive, I haven't signed up yet. However, More told me that life insurance can help provide financial means to get the cryonics procedure done. Now I'm set on signing up for cryonics before my presidential campaign ends, in hopes of bringing attention to this small but potentially life-changing industry.

When I asked More why more people don't sign up for cryonics, he shrugged and said, "I don't really know. You would think everyone that likes living would be interested in this. But so far few people have signed up."

As an aspiring politician, I advocate for government policy that specifically protects citizens' lifespans. Today, cryonics is the only hands-on treatment I know of that has a shot a preserving the lives and minds of the people we love. If people could become more comfortable with the idea of cryonics as emergency medicine rather than simply "freezing the dead," then I think it might become a much larger industry.

.

****

36) Atheist Jacque Fresco: Eliminating Money, Taxes, and Ownership Will Bring Forth Technoutopia

Futurist and architect Jacque Fresco speaks in parables. If he goes on too long with a story, his 40-year partner Roxanne Meadows interjects facts to keep him on track. Fresco recently turned 100 years old, and is the oldest celebrity futurist in the world. His magnum opus is The Venus Project, a 21-acre Central Florida Eden with white dome-shaped buildings that Meadows and he hand built over three and a half decades. The sanctuary and research center is where Fresco still leads weekly seminars, which includes a tour of 10 buildings—some filled with hundreds of future city models inside them—that highlight the promise of a future world where equality and technology abound.

How I met Fresco at The Venus Project this month starts with income taxes—something I hate and aim to one day eliminate altogether for humanity. Fresco doesn't like taxes either. While searching online about taxes, I stumbled upon Fresco's voluminous work: over 80 years of essays, filmed lectures, books, documentaries, models, and architectural drawings. Much of Fresco's work is anchored by his main philosophical idea: a resource-based economy, where there's not only zero taxes, but no ownership or money either.

It sounds fanciful, but the more I read about Fresco's work and ideas, the more intrigued I became. Here was a man with a vision, one not dissimilar from my own. The timing of my meeting with Fresco and Meadows was serendipitous. As I neared the end of my US presidential campaign, I was looking to build out the Transhumanist Party's 20-point platform with a more aggressive futurist platform—one that looked not only 10-20 years into the future, as I generally focus on, but one that also examines what could and should happen in 50 years or even the next century.

Over the next 20 years, I see automation taking nearly all jobs, and I doubt capitalism will survive that. As a result, I advocate for beginning the process of eliminating taxes and doling out a universal basic income—one that swallows welfare, Social Security, and all health services. Otherwise, I see inequality dramatically growing and an even larger befuddled welfare system than we have now. When robots take all the jobs, I also see civil strife and revolution occurring if corporations and the government don't give back enough to society.

For me, the most important aspect of the future is to actually get there, and I worry that without giving something to unemployed humans, a dystopic society of violence and chaos will come about. The last thing America—and the scientific community—needs is a civil war.

Some experts have predicted that fully automated luxury communism is the way to go, and it's a term increasingly being thrown around. Basically, it argues that humans should be pampered by technology, and to do so, communism should finally become the dominant economic system. Fresco doesn't buy this.

He thinks that if we could just get rid of money and ownership, most of the humanity's problems would disappear. And he claims only a resource-based economy—an idea he said he's been working on since he was 13 years old—could do this.

The resource-based economy goes like this: In the future robots will do all the jobs (including creating new robots and fixing broken one). Now, imagine the world is like a public library, where you can borrow any book you want but never own it. Fresco wants all enterprise like this, whether it's groceries, new tech, gasoline, or alcohol. He wants everything free and eventually provided to us by robots, software, and automation.

Fresco's system is indeed beyond capitalism or communism, which he says are economic systems based upon scarcity. Fresco wants a system that is based on abundance. He thinks the world's resources could provide for humanity many times over—something I also agree with. In fact, my plan for funding a universal basic income partially comes from utilizing America's untapped government resources; half of the land in our 11 most western states belongs to the government, resulting in trillions of dollars of untapped wealth.

Despite being over a century old, Fresco is still mentally sharp. He does speak noticeably slowly, but he's quite capable of arguing for his resource-based economy. Listening to him, it's impossible to not feel like he is a deeply spiritual man. In fact, his rhetoric is borderline religious-sounding. He thinks money is the root of all evil.

"If I meet a Christian," Fresco said, "I ask him: Why should we have ownership of property on Earth when there is none in Heaven?"

But Fresco is no believer. Like me, he is an adamant atheist. And also like me and millions of others around the world, he thinks all the world's problems can be solved by reason and using the scientific method.

I find it ironic that Fresco's ideas are Christ-like. He supports an open system where everything is free and equal. He even supports open borders, something I've advocated too as the world gets more interconnected and countries merge—despite the recent success of Brexit.

The real ice breaker question was asked by an entrepreneur on The Venus Project tour to Roxanne: "This all sounds very interesting. But how do we actually transition to this new society and way of living?"

A resource-based economy, where one day no one will work but everyone has plenty, might just be the trick to gaining piece of Heaven on Earth.

Meadows answered saying, "We need to build one city to show the world how this could work. If we could just build one futurist city and show everyone how well this system works, people would demand it around the world."

Finland recently began an experiment with universal basic income. And across the continents, other communities run various forms of alternative government, such as the Mareki community on the island of Vanuatu, which doesn't use money or have ownership—everything is community based. Additionally, the Seasteading Institute, of which I am an ambassador, is trying to create the first truly Libertarian nation.

So creating new governments, societies, or cities is possible. Fresco—who recently won an award from the United Nations—has designed numerous cities that he wants his first experimental city to look and be like.

Even though some of the ideal robot and automation technology may not be here yet, Meadows still believes the city could be started now and be successful.

She wrote me, "This city does not depend on the predictions of automation for the near future. Most aspects of it can be accomplished now even though it would not be fully automated."

Additionally, costs for such a city could be less than normal. Jennifer Huse, the social media director at The Venus Project, told me, "Fresco's city designs would be less expensive than any others to build because you only have to design one eighth of the city—then duplicate it."

Some of Fresco's architectural city designs are known for easily connecting into each other and coming from the same mold, thereby being easier, cheaper, and simpler to create.

I advocate for such a city. How wonderful and exciting that would be to visit. No one knows if a city like this will work yet—and of course, for a complete resource-based economy to function smoothly, it would require having the world on board and participating. Nonetheless, I believe in trying and setting a powerful example for the future.

Something is rotten with the world and how people treat each other. We've undergone wars, crime, enslavement, and poverty for too long. Fresco's ideas are a breath of fresh air to grant equality and prosperity to all people—and to get us all living a better, more interesting life. A resource-based economy, where one day no one will work but everyone has plenty, might just be the trick to gaining piece of Heaven on Earth.

CHAPTER VII: OTHER GODLESS ESSAYS

37) For Christians, Does Being Pro-Life Lead More Souls to Hell?

In late November, the Colorado Planned Parenthood shooting where three people were killed and nine wounded sadly reminded Americans again that women are not safe in this nation when trying to make choices about their bodies. It compelled me to take a candid spiritual look at the popular Christian stance on abortion.

As a transhumanist US Presidential candidate, I am pro-choice. More interestingly, though, as a former Christian and Catholic school student, I was also pro-choice.

Here's why: From a strictly biblical point of view, being born on Earth is a test. All our actions will eventually be judged by an omnipotent God who will determine whether we go to heaven or hell. Our deeds, sinful or not, determine where we spend eternity. The Bible even says many people will not get into heaven because it's quite difficult to be a sin-free Christian—meaning the majority of human souls may spend an endless amount of days in hell. Like it or not, about 2.2 billion people on Earth believe in these ideas. Another 1.6 billion Muslims believe in mostly the same thing, too.

Where the Abrahamic rabbit hole gets weird—at least for me—is the fact that many Christians (and Muslims) believe that an aborted fetus goes to heaven.

The metaphysical impact of that religious belief is just bizarre. It means that the most sure thing to do to get a soul into heaven is to abort a fetus before it leaves its mother's womb and has the chance to sin. The crazy thing is this makes abortion providers some of the most considerate, humanitarian people we know—at least from an Abrahamic religious perspective.

Abortion providers and pro-choice advocates have long been filling heaven with pure souls—instead of committing them to a lifetime on Earth, challenged with trials of sex, drugs, and transhumanism.

Of course, the logic described above certainly sours the pro-life argument that abortion is evil. However, the premise of that logic is anything but certain or sound. It requires basing views on leaps of faith and ancient religious mores—like the existence of God, hell, and heaven. The reality—despite the billions who believe in formal religion—is that no one really knows anything definitely. While I like to say I'm atheist just to be defiant in the face of an overly religious civilization, like any sensible person, I honestly don't know what exists beyond me and the material universe I live in.

The fact is—since we're all empathetic mammals—that no one likes to have abortions, and no one likes to provide them. In an ideal world, no one would ever get pregnant unless they were certain they were ready for parenthood and knew they were going to have a perfect child. But stuff happens and things don't go as planned, and people must do things to try to accomplish the greater good for themselves and society. In this way, we should be grateful that women's clinics—like Planned Parenthood—are there to help out so that the best choice about one's life can be made.

And if we insist on being a Christian, then we might want to look at the bigger spiritual picture and stop trying to shoot people that are helping a soul's entrance into heaven.

****

38) A World Future Society Conference Speech: Everyone Faces a Transhumanist Wager

Recently, I had the honor to give a speech at the World Futurist Society's conference in Orlando, Florida. The World Futurist Society is the largest nonprofit organization of its kind with over 25,000 members in nearly 100 countries. Its yearly conference is a mecca for thousands of futurists looking to hear the latest forward-looking news and ideas. Hundreds of speeches, workshops, panels, meet-the-author sessions, poster presentations, and luncheons occurred.

My speech at the conference was loosely based on an essay I recently wrote titled *Everyone Faces a Transhumanist Wager*. I wanted to share a condensed version of the talk because it presents a fundamental dilemma every human being on the planet must confront. Here's the shortened speech:

Ladies and gentlemen, we have a problem. Each one of us has a problem. In fact, no matter where you go on the planet, no matter who you find, every single person on Earth has this same dire problem.

That problem is our mortality. That problem is called death.

The reason it's a problem is because human beings love life. We all love the precious chance of existence. Even in one's darkest psychological despair, or one's most exhausting hardship, or one's most catastrophic tragedy, the thing we call life is still always miraculous. We cherish life and we don't want to lose it or have it end.

But end it will. No matter how much we wish otherwise. The stark truth is right before our eyes—that nothing in today's world can save us from death. The obviousness of this overwhelms us every time we see a loved one or a friend whose body is lifeless, never to reach out, touch, and communicate with us again. Death is final.

The great irony for our species is that we don't just have this one problem—but two problems. The second problem is nearly as vicious as the first. The second problem is the fact that most people around the world are just not worried about the first problem—they're not worried about dying. They're either religious and have the supposed afterlife all worked out, or they just don't care, or they just don't think conquering human death is possible. Whatever people's reasons, they just don't see the first problem as serious enough to warrant immediate concern—especially in a meaningful, tangible way that makes them not die. And by not recognizing death as a problem, many people have no reason to attempt to defeat it.

I have made it a mission in my life to make people aware of these two problems. It is why I wrote my philosophical novel *The Transhumanist Wager*. The concept of the Transhumanist Wager in the book is simple. It explains that in the 21st Century—the age of unprecedented technological innovation—it is a betrayal of ourselves (and the potential of our best selves) to not tackle and solve our two most pressing problems using modern science. More importantly, my book explains how we can solve these two problems.

But first, some of you are asking: What is a transhumanist? What does such a person want? What are the main goals? Some people around the world still don't know what transhumanism means. When explaining the term to people, I find it easiest to use the Latin translation. "Transhuman" literally means beyond human.

Transhumanist goals are broad and varied, but mostly they revolve around human beings using science and technology to radically improve and enhance themselves, their lives, and society. Transhumanists often concentrate on stopping or reversing aging—we are sometimes called life-extensionists or longevity advocates. Many transhumanists also focus on robotics, bionics, artificial intelligence, biohacking, and other similar fields of study. Transhumanists are often, but not always, nonreligious. They find meaning in their own lives and possibilities, without a divine creator. The philosophies of transhumanism make it possible that in the future—using extreme science and technology—one may become a so-called divine creator if they wanted. In almost all circumstances, transhumanists prefer reason over any other method of understanding to guide themselves in life.

Every transhumanist comes to their own realization of why they feel they are a transhumanist. Each path is unique, personal, and totally different than another. I want to tell you briefly about my path. I was first introduced to transhumanism as a philosophy student attending Columbia University in New York City. For a class assignment, I was told to read a magazine article on some of the recent breakthroughs in cryonics. The article described a small but passionate group of scientists who believed that science and technology would be able to bring frozen patients back to life in the future if they were preserved properly. The article also discussed the transhumanism movement, which it described as a community of reason-based futurists who wanted to use science and technology to improve their lives and live indefinitely. I was deeply intrigued. I finished that article and wanted to know more. I spent the next ten years reading everything I could find on future technologies, human enhancement, and transhumanism.

However, it wasn't until I was in the jungles of the demilitarized zone of Vietnam as a journalist for the National Geographic Channel that I came to dedicate my life to the field of transhumanism—that I came to the powerful conviction that human life should be preserved indefinitely. While in the jungle filming Vietnamese bomb diggers searching the ground for unexploded ordinances to recover and sell, I almost stepped on a partially unburied landmine. My guide pushed me out of the way, and I fell to within a foot of the mine. Tens of thousands have died from landmines in the DMZ in the last forty years, and I was lucky I was not one of them.

For me, nothing was ever the same again after that moment. The landmine incident permanently stamped into my mind how fragile the human body was—how precious our minutes alive on this planet really are. Upon returning to the Unites States, I began writing *The Transhumanist Wager*. The reason I tell you my personal story about becoming a transhumanist is that every one of us has their own story. But the two main problems we each face: death, and general apathy of death—and the choice we must make regarding them: a Transhumanist Wager—that is not just for some people. It is for every reasonable person in the world.

Indeed, in the quickly advancing 21st Century, making a Transhumanist Wager approaches us now as an ultimatum—the most challenging one we may ever face. Luckily, given how fast modern science is growing and changing our lives, making the wager is also the only reasonable option. If you love life, you will dedicate yourself to finding a way to preserve that life. Transhumanists do not want to preserve their life via heaven-promising religions, false hopes, an unconscious mystic super spirituality, or otherwise. There are only rational ways transhumanists will do it: through the tools they can create with their own hands; through the reason their brains can muster; and through the conviction their being prompts of them by not wanting to die and disappear. To do otherwise in today's world is to remain irrational and, as my novel discusses, to be masochistic and even borderline suicidal. In a world where we have the technology to travel to Mars, where we can video chat on our cell phones to someone 10,000 miles away, or we can triple the lifespan of mice with biotechnology, it's our evolutionary destiny to significantly extend our lives and to be transhuman.

Once you have identified the human race's two main problems, and you understand that you each face a Transhumanist Wager, the question is: what to do? How can you solve these problems and make the right choice in the wager.

It's quite simple, really. The journey of the transhumanist requires no ritual, no prayer, and no spiritual sacrifice or payment. It requires only your ability to reason. Ask yourself how you can best dedicate yourself to a specific cause of transhumanism and its various fields: aging research, cyborgology, stem cell science, suspended animation, singularitarianism, genetic engineering, machine intelligence, or the dozens of other areas. Then do it. For some, this may mean going into science or technology as a new career. For others it will mean volunteering in transhuman groups that need support. For some it will mean going into politics and pushing for more science-friendly laws. For others, it will mean donating resources to scientific centers and struggling innovators. For some, it will mean creating transhumanist art and using it a vehicle to push for a more scientific-minded society. For others it will mean just talking with friends and family about why you think science and technology are the best drivers of civilization.

Whatever it is that one can do, be transhumanist-minded. Be a people that belongs to a bright, rational scientific future, not one dogged by the old ways of archaic institutions, apathy, fear, or primitivism. Be transhuman, and let us all embrace our evolutionary destiny and the joys of perfect health and being that science can help us reach.

****

39) Why Haven't We Met Aliens Yet? Because They've Evolved into AI

While traveling in Western Samoa many years ago, I met a young Harvard University graduate student researching ants. He invited me on a hike into the jungles to assist with his search for the tiny insect. He told me his goal was to discover a new species of ant, in hopes it might be named after him one day.

Whenever I look up at the stars at night pondering the cosmos, I think of my ant collector friend, kneeling in the jungle with a magnifying glass, scouring the earth. I think of him, because I believe in aliens—and I've often wondered if aliens are doing the same to us.

Believing in aliens—or insanely smart artificial intelligences existing in the universe—has become very fashionable in the last 10 years. And discussing its central dilemma: the Fermi paradox, has become even more so. The Fermi paradox states that the universe is very big—with maybe a trillion galaxies that might contain 500 billion stars and planets each—and out of that insanely large number, it would only take a tiny fraction of them to have habitable planets capable of bringing forth life.

Whatever you think, the numbers point to the insane fact that aliens don't just exist, but probably billions of species of aliens exist. And the Fermi paradox asks: With so many alien civilizations out there, why haven't we found them? Or why haven't they found us?

The Fermi paradox's Wikipedia page has dozens of answers about why we haven't heard from superintelligent aliens, ranging from "it is too expensive to spread physically throughout the galaxy" to "intelligent civilizations are too far apart in space or time" to crazy talk like "it is the nature of intelligent life to destroy itself."

Given that our planet is only 4.5 billion years old in a universe that many experts think is pushing 14 billion years, it's safe to say most aliens are way smarter than us. After all, with intelligence, there is a massive divide between the quality of intelligences. There's ant level intelligence. There's human intelligence. And then there's the hypothetical intelligence of aliens—presumably ones who have reached the singularity.

The singularity, Kevin Kelly, co-founder of *Wired Magazine*, says, is the point at which "all the change in the last million years will be superseded by the change in the next five minutes."

If Kelly is correct about how fast the singularity accelerates change—and I think he is—in all probability, many alien species will be trillions of times more intelligent than people.

Put yourself in the shoes of extraterrestrial intelligence and consider what that means. If you were a trillion times smarter than a human being, would you notice the human race at all? Or if you did, would you care? After all, do you notice the 100 trillion microbes or more in your body? No, unless they happen to give you health problems, like E. coli and other sicknesses. More on that later.

One of the big problems with our understandings of aliens has to do with Hollywood. Movies and television have led us to think of aliens as green, slimy creatures traveling around in flying saucers. Nonsense. I think if advanced aliens have just 250 years more evolution than us, they almost certainly won't be static physical beings anymore—at least not in the molecular sense. They also won't be artificial intelligences living in machines either, which is what I believe humans are evolving into this century. No, becoming machine intelligence is just another passing phase of evolution—one that might only last a few decades for humans, if that.

Truly advanced intelligence will likely be organized intelligently on the atomic scale, and likely even on scales far smaller. Aliens will evolve until they are pure, willful conscious energy— and maybe even something beyond that. They long ago realized that biology and ones and zeroes in machines was literally too rudimentary to be very functional. True advanced intelligence will be spirit-like—maybe even on par with some people's ideas of ghosts.

On a long enough time horizon, every biological species would at some point evolve into machines, and then evolve into intelligent energy with a consciousness. Such brilliant life might have the ability to span millions of lights years nearly instantaneously throughout the universe, morphing into whatever form it wanted.

Like all evolving life, the key to attaining the highest form of being and intelligence possible was to intimately become and control the best universal elements—those that are conducive to such goals, especially personal power over nature. Everything else in advanced alien evolution is discarded as nonfunctional and nonessential.

All intelligence in the universe, like all matter and energy, follows patterns—based on rules of physics. We engage—and often battle—those patterns and rules, until we understand them, and utilize them as best as possible. Such is evolution. And the universe is imbued with wanting life to arise and evolve, as MIT physicist Jeremy England, points out in his *Quanta Magazine* article titled *A New Physics Theory of Life*.

Back to my ant collector friend in Western Samoa. It would be nice to believe that the difference between the ant collector and the ant's intelligence was the same between humans and very sophisticated aliens. Sadly, that is not the case. Not even close.

The difference between a species that has just 100 more years of evolution than us could be a billion times that of an ant versus a human—given the acceleration of intelligence. Now consider an added billion years of evolution. This is way beyond comparing apples and oranges.

The crux of the problem with aliens and humans is we're not hearing or seeing them because we don't have ways to understand their language. It's simply beyond our comprehension and physical abilities. Millions of singularities have already happened, but we're similar to blind bacteria in our bodies running around cluelessly.

The good news, though, is we're about to make contact with the best of the aliens out there. Or rather they're about to school us. The reason: The universe is precious, and in approximately a century's time, humans may be able to conduct physics experiments that could level the entire universe—such as building massive particle accelerators that make the God particle swallow the cosmos whole.

Like a grumpy landlord at the door, alien intelligence will make contact and let us know what we can and can't do when it comes to messing with the real estate of the universe. Knock. Knock.

****

40) Why I Advocate for Becoming a Machine

Transhumanists want to use technology and science to become more than human. Naturally, in this process, certain elements of our humanness will be replaced and likely lost. Many people have conflicting feelings about this. I don't.

Part of the problem with people's perceptions of losing their humanness is not their fear of becoming something else, but their inability to empathize with their future selves. I want people to know their future transhuman self is almost certainly going to be more amazing, beautiful, and unique than their current self.

To understand why one's future self could be markedly improved over one's current self, consider how we perceive reality for a moment. Human beings have five basic senses that send signals to our brain, telling us what's out there in the world. These senses understand only tiny bits of the universe around us. For example, our eyes can only see about 1 percent of the light spectrum. Our ears aren't much better: they are unable to register many noises that other animals like dogs, dolphins, and bats can hear. Our sense of touch basically only works if we're actually touching something.

Despite all these obvious physical inabilities, humans insist what we experience is "reality." However, reality to someone with built-in microscopic or telephoto vision and hyper-sensitive hearing is potentially many times more complex and profound than anything a natural human being might experience.

Around us are many things we never notice, like energy patterns that have traversed the universe over thousands of light years, or sounds waves from whales across the ocean, or vibrations that started from the core of the Earth. But we humans are oblivious, unless, of course, we are in a laboratory somewhere and happen to be studying this phenomena with specialized scientific machinery.

In the near future, however, these abilities to pick up on the greater essence and profundities of the universe will be standard equipment for cyborgs. Already, robotic eyes in blind patients offer telescopic possibilities that no human eye can match. Some Cochlear implants for the deaf also can pick up normally inaudible noises to the natural human ear. And touching, smelling, and tasting can all be improved using a variety of different types of advanced tech sensors.

When we individually replace or augment a human body part—such as giving someone an artificial hip—most people don't see that as becoming a cyborg. Additionally, it really doesn't matter if the part replaced or improved is a heart with a robotic pump, or a knee with a titanium joint, or a penis with a built-in balloon for help stiffening—all technologies which already exist. We usually think of such transformation as needed medical treatment, or even elective vanity surgery in some cases.

However, if someone was to get 10 transhumanists upgrades for their body all at once, then the flavor of what a person has become gets downright dystopian in many people's minds. Many in the public would now say that person is something not quite human anymore. They'd also surely think that person is a weirdo.

But of course, nothing is wrong with 10 bodily changes at once—or 50 for that matter. That installed robotic heart allows you to have much better, long lasting sex. And that artificial knee will allow you to get your tennis game back. And that part-synthetic sexual organ might be the beginning of many new adventures.

This fine line between transhumanist upgrades and what makes us uncomfortable about too much technology in our bodies is a bizarre psychological conundrum. It's so challenging that I believe the next great civil rights debate around the world will be about how much humans should embrace radical technologies in their bodies—and when they should just say "no" to upgrades.

The good news is I think most people would agree that even replacing most every inner organ in your body is not becoming a cyborg or something machine-like.

But mess too much with the outer body, and everything changes quickly. When we propose electively replacing limbs, for example, most people feel something has fundamentally changed in the human being. A line has been crossed that cannot easily be undone. We may still have a mind of flesh, but our eyes tell us we are now partially a machine and something very different than before. And that freaks people out.

It really shouldn't, though. The benefits are obvious for artificial limbs, such as indefinite durability, ease of upgrades, and immunity to skin cancer or even snake bites (which kill 45,000 people in India alone every year).

What's not so obvious is how humans can become psychologically comfortable with their growing cyborg identity. Unfortunately, the whole process is going to be an uphill battle. Hollywood seems intent on insisting that humans must fight and win against machines, not join with them. Formal Abrahamic religion insists we should not strive to be gods and that dying is good since that's the way to meet God in heaven.

To better adjust to our coming transhuman form and our merging with synthetic parts, we need new, positive messages that the cyborg era is not the end of the human age, but the expansion of it. The same can be said of machines, which we also will one day become.

The reality is that many transhumanists want to change themselves dramatically. They want to replace limbs with mechanical endoskeleton parts so they can throw a football further than a mile. They want to bench press over a ton of weight. They want their metal fingertips to know the exact temperature of their coffee. In fact, they even want to warm or cool down their coffee with a finger tip, which will likely have a heating and cooling function embedded in it.

Biology is simply not the best system out there for our species' evolution. It's frail, terminal, and needs to be upgraded. In fact, even machines may be upgraded in the future too, and rendered as junk as our intelligences figure out ways to become beings of pure conscious energy. "Onward" is the classic transhumanist mantra.

No matter what happens, to move forward in the transhumanist age, we need to let go of our egos and our shallow sense of identity; in short, we need to get over ourselves. The permanence of our species lies in our ability to reason, think, and remember who we are and where we've been. The rest is just an impermanent shell that changes—and it has already been changing for tens of millions of years in the form of sentient evolution.

****

41) A Brain Implant that Registers Trauma Could Help Prevent Rape, Tragedy, and Crime—So Why Don't We Have It Yet?

There's been a lot of talk across America about college rape culture in the last few weeks. Much major media has been highlighting the persistent and unfortunate problem. Perhaps the most well known article came from *Rolling Stone*, which ran a controversial story highlighting seven University of Virginia fraternity students allegedly raping one freshman girl for hours during a party. In the wake of so much arresting coverage, numerous universities and legislative bodies are considering new methods to deal with the problem.

So far, those new methods seem to consist mostly of advocating for clearer language to stop the violence from happening in the first place and greater transparency in the rape victim's reporting process. I'm not optimistic the changes will do much to stop rape and other forms of criminal violence in any significant way. There are too many aggressive, idiotic men out there—and yes, men are almost always responsible for the violence.

The facts of domestic abuse in America are sobering. Nonprofit Arkansas Coalition Against Domestic Violence reports that every 15 seconds a woman is beaten and that 35 percent of all emergency room visits are a result of domestic violence. Nonprofit A.A.R.D.V.A.R.C., An Abuse, Rape, and Domestic Violence Aid and Resource Collection, reports that that the US Surgeon General states that domestic violence is the leading cause of injury to women between the ages of 15 and 44 in the United States. According to *Feminist.com*, over 22 million women in the United States have been raped in their lifetime, based on the National Intimate Partner and Sexual Violence Survey 2010. SafeHorizon, the largest victims' service agency in America, says there are 2.9 million reports of child abuse every year nationally, and it costs $124 billion dollars annually in medical, court, law enforcement, child welfare, and juvenile protection services to deal with the problem.

As a transhumanist, I strive to consider all societal problems from technological and scientific points of views. It turns out there might be a simple solution that could reduce rape and some violent crime all across the country. I call it the trauma alert implant.

Cranial implants and brain wave technology—despite a Mark of the Beast reputation by Christian conspiracy theorists—have come a long way in the last few years. Already, hundreds of thousands of people in the world have microchip implants in their heads, consisting of everything from chips to help Parkinson's sufferers to cochlear implants for the deaf to devices to assist Alzheimer's patients with memory loss. For each, this technology allows a better life. DARPA recently announced a $70 million dollar five-year plan to develop implants that can monitor soldier's health. It's part of President Obama's new multi-billion dollar BRAIN Initiative.

Implants using Electroencephalography (EEG) technology can read and decipher brain waves. Trauma, however experienced, is a measurable biological phenomenon that can be monitored and captured by an implant device. Scientists must do nothing more than create a trackable chip that sends an emergency signal to nearby authorities when it registers extreme trauma. Help can then arrive quickly to the victim.

Much of the technology for such a device basically already exists. And such a device could be useful for far more than rape or criminal violence, too. Drowning, being burned in a fire, automobile accidents, building collapses, snake bites, kidnappings, bullet wounds, senior citizens who've fallen down stairs and can't get up—the list of terrible things that happen to humans goes on and on. The result of every one of them is almost always the same: brain waves that manifest extreme trauma—the human's most basic response and alert system. Regardless what misfortune happens to a human being, most experts agree that getting victims rapid emergency assistance is the single best way to help them.

Consider the 2-year-old boy was snatched away from its parents by an alligator at Walt Disney World in 2016. I have a similar-aged toddler myself, and I followed this heartbreaking story closely. Unfortunately, it ended as horribly as it began, with the recovery of a dead child.

While scene reports claim the father got into the water to save his son, perhaps if that 2-year-old at Disney World had been GPS chipped, the parents could have tracked him on their smartphones. And security might have been able to quickly identify his location in the water, perhaps even fast enough to have rescued him. *The New York Times* reports that the body was tragically found underwater only 10-15 feet from where it was last seen.

That's of little consolation now, of course, and I don't mean to be insensitive to the family's loss, but I do think this tragedy illustrates how implants could help improve public safety. They could help track our children, and adults for that matter, in the case of kidnapping and Amber alerts, or even just when they get lost on a hike in the woods.

As the father of a 5-year-old who will be attending school next year, I'm a big believer in the future that all children will get chipped somewhere on their bodies, perhaps like all children get vaccines in the U.S. It's crazy to me that we don't develop and use this technology, especially with our children. I'm looking into getting my children chipped after this alligator incident and because, as a controversial presidential candidate, I have security issues myself to worry about.

Of course, it's not only implants. It's chip tattoos, GPS jewelry, wearable tech T-shirts, or even shoes with tracking tech built into them. Using tech to keep humanity safe is a burgeoning field. Interestingly, an industry already exists around children using tech to keep their whereabouts safe, but they're mostly children with disabilities—some who have a propensity to wander off.

Perhaps the most advanced case of chipping people already in existence has to do with the military. Reports describe special forces experimenting with them so they can be tracked. In 2014, for example, the U.S. Department of Defense announced a $26 million grant for a brain implant that would record, analyze, and potentially alter live electrical signals to soldiers. The military is getting so interested in implants that I was recently asked to consult the U.S. Navy on research of chipping their service people.

Back to the trauma alert implant—which to me is the holy grail of safety. Another great thing about it is that not everyone would have to get it to help stop violent crime and domestic abuse. In fact, probably most people wouldn't (though, I surmise in the future many people will get one for a multitude of reasons). The existence of the chip itself—similar to a possible hidden camera in a room—would be enough to scare off many criminals, who would always be second guessing if their victim had one. This would especially be the case when it comes to crimes that are hard to prove or go habitually unreported, such as date rape.

All things considered, the trauma alert implant sounds like a sensible and impressive thing. So why don't we have them yet?

To begin with, Americans are wary of brain implants. They don't mind holding a cell phone to their ears for a half hour, but ask them to get a piece of sophisticated tech inside their heads and many freak out. They squawk how weird it is and that they don't want to be a cyborg (all the while spending untold hours surfing the internet, flying on jet airplanes 30,000 feet in the air, and taking multiple vaccines and pills). The transhumanist age is already here, whether it's weird or not. For most people, it's just a matter of culturally accepting it.

Another complaint that people have with implants is the privacy issue. Nobody wants to be trackable. Sure, that's understandable. But bear in mind, that every time you get on the internet, stop at a gas station, or use your credit card, you're already being tracked. We may all distrust surveillance, but that's not going to stop the gargantuan amount of cameras recording in America right now, many of them in public places. While solid information is hard to find on how many cameras are operating in America, Wikipedia reports that Chicago has at least 10,000, and the United Kingdom may have as many 4.2 million cameras, or 1 for every 14 people. The good news is, just like our cell phones we carry around, we'll likely have the option to turn off our implants anytime we want, thereby giving us control of who can watch us. Additionally, surely implants could be programmed so that they could "only" be tracked once they were triggered for extreme trauma.

But probably the most significant reason Americans dislike implants is because of religion. At least 80 percent of the country's population holds some form of faith—mostly of Christian denomination. And a significant number of Christian people consider brain implant technology to be the definitive Mark of the Beast—and a sign of End Times. I'm an agnostic tending towards atheism, so I don't understand those fears. But I do know that Revelations and the Second Coming of Jesus supposedly can't be stopped by people, according to the Bible, so perhaps Americans should work through their cyborg-phobias and embrace useful transhumanist technology. After all, if a gang of rape perpetrators suspected their victim had a trauma alert chip that would notify authorities, do you think they'd still commit the crime? Surely most wouldn't, especially not university students.

I'm grateful my young daughters will probably never have to worry about drunk drivers on their prom night. In a decade's time, most cars on the road in America will be driverless or come with alcohol detection systems that don't allow inebriated drivers. Such technological innovation is just a drop in the ocean of the benefits that progress brings to our world. The truth is that technology can help fix almost all the world's problems. It can also help with the tragedy of rape and criminal violence, which dramatically harms nearly 10,000 women and children a day in the US. If even a small portion of the population would have trauma alert implants, rape and criminal violence might be substantially reduced.

****

42) Glass Coffins: We Are in a Race to Save Lives, But Religion Wants to Take Them

Easter celebrations. The swearing in of President Obama with a bible. The Pledge of Allegiance's "One nation under God." Ted Cruz's freakish evangelism. Even when we sneeze, many Americans still reflexively say: God bless you.

American society exists in an engulfing religious framework— and that inescapable Abrahamic point of view leads to one ultimate goal: an eternal afterlife with the maker. For the majority of Americans—about 70 percent—that is what they believe and how they live in this world.

Whether you like it not, America is a culture mired in deathism—the idea that human death is natural, inescapable, and ultimately desirable because it unites one with God.

Lately, though, the burgeoning transhumanism movement is challenging all that. Gerontologists, crynocists, singularitarians, biohackers, roboticists, geneticists, futurists, and anti-aging activists—all considered part of the transhumanist platform— are standing up and demanding humans conquer biological death.

Once seen as fringe, but now increasingly seen as potentially visionary, transhumanists are challenging the very nature of what it means to be a human being. But they're also being chained down with obsessive laws and government regulations that hinder their research and aims to keep all humans alive. The Food and Drug Administration (FDA), which approves radical transhumanist drugs, medical devices, and gene therapies is one of the slowest, most bureaucratic medical-approving bodies in the world—besides generally being run by Christians. Some US medical companies get so bogged down by the FDA, that transhumanists CEOs take their research out of the country to move forward, even if that's controversial and unconventional to the public.

Viewers need to understand that transhumanists believe they are fighting for our lives. Similar to how the HIV-stricken community and its citizen scientists battled the AIDS epidemic in the 1980s when the US government and big pharma companies wouldn't help—made famous by the documentary *How to Survive a Plague*—transhumanists are literally battling death right now.

I speak for most transhumanists when I say we don't have time to follow all the damn laws, regulations, and endless Federal recommendations of the bloated government when 150,000 people a day are dying in the world—and many science experts in the world believe we will conquer death in this century.

We are in a desperate race to find those life extension technologies and save hundreds of millions of lives. And if transhumanists appear to sometimes make decisions as if they were living in the Wild West—it's because we are involved in the most important crusade humanity has ever faced. There's no time to lose. We are in a war against death. The difference between humanity overcoming death in 2030 versus the year 2050 is the difference of saving 1 billion human lives. Chew on that for a moment. This is a race.

If I weren't an atheist and believed in afterlives like most people in America, I'd be a very different person. I wouldn't need to be a transhumanist, because death would just be a transition into another living realm. But transhumanists don't feel that way. Death means the loss of consciousness forever and the physical cellular breakdown of the body. It means never ever having another thought, and becoming dirt and food for bugs.

Ask nearly any transhumanist what they most want out of life, and they will tell you they want to escape the expiration date of their existence. And they'll also tell you that government regulations, the US military, and the FDA are simply part of an overarching culture that indirectly facilitates the death of human beings—especially since the US Congress, the President, and all Supreme Court members believe in an afterlife. If America is not a culture of monopolistic fundamental religiosity and fatalism, then I don't know what is.

Realizing you don't want to die, and that you also don't believe in an afterlife, usually takes a personal epiphany. Every transhumanist has a conversion moment—an instant when they make a Transhumanist Wager and decide dying is not acceptable. Mine happened when I almost stepped on a landmine in Vietnam while working for National Geographic. Afterward, I thought: What the hell? I almost died and had my body blown in half. Death is the stupidest and most tragic experience ever. Something must be done about it.

Sadly, it's almost impossible to tell people the follies of dying—even if they love life. Every single person must arrive at this understanding for themselves. It's a very personal revelation, and not one that can be taught or forced. Becoming a transhumanist—becoming dedicated to overcoming death with science and technology—must be experienced personally in a heartfelt and deeply philosophical way.

That said, an entire international science movement has now sprung up to help this process along of wanting to live indefinitely—to defy the culture of death we are all surrounded by. Cryogenic facilities are out there like Alcor and the Cryonics Institute. Nonprofits like SENS Research Foundation do amazing work. Singularity University educates young science and life extension entrepreneurs. Private companies like Insilico Medicine which uses big data to figure out how live longer now exist. And of course, communities like the Church of Perpetual Life have recently sprung up. Even what is being billed as the largest life extension gathering of its kind, the RAAD Festival, is happening this August in San Diego—with a hoped-for 2,000 participants.

To wage war against death, transhumanists are starting new life extension enterprises more quickly than ever before. Some of these projects and organizations are brilliant and some are wacky—and some surely will fail and disappear. But the more partners and allies transhumanism has, the better. And greater chances of overall success will come from putting our hopes into visionaries like Bill Falloon, Martine Rothblatt, Dr. Aubrey de Grey, and the many other controversial and colorful figures in the life extension movement.

The alternative is putting our hope into someone like the Pope, who maintains that condoms are sinful—and the result is that millions of people in Africa will likely die from AIDS as a result because they contract HIV. Or in George W. Bush, who stopped life-saving stem cell research for seven years in America because of religious beliefs. Or in presidential candidate Ted Cruz, who makes decisions about science and medical policy based on a 3,000-year-old book that says Jesus needs to forgive us our sins so we can live with him forever in bliss. Pure insanity. (But very useful for the power hungry conservatives who use religion to control America's moral, cultural, and philosophical outlook.)

Sometimes I think we ought to have glass coffins for our loved ones who die, and keep them in viewing rooms, so we can watch them decompose—so we can see all that we loved about people close to us slowly disappear into nothingness. It would force us to realize something quintessential—that if we're wrong about the afterlife, we forever lose this precious miracle called life.

If America cares about the health of its people, it needs to join the war on conquering aging and death. Many gerontologists believe we could stop or reverse aging within 25 years if enough funding is put into it. With that in mind, America needs to recognize that in the twenty-first century, dying is a public safety issue—that people deserve a universal right to indefinite lifespans, as put forth in the Transhumanist Bill of Rights. And the government must take a leading role in establishing the science and forming the culture to offer people their maximum longevity—in the same way it offers vaccinations to avoid disease.

Thankfully, hundreds of thousands of people around America are starting to get it. Google's life extension company Calico was recently formed to battle aging. Facebook's Mark Zuckerberg's recently made positive statements about trying to cure disease and live much longer. Billionaire Peter Thiel generously supports life extension research. And Ray Kurzweil's dedication to overcoming involuntary death is in the media all the time. It's all proof that some of the best minds of the world are joining the battle to conquer death.

When people look back in 50 years—and we all hopefully have the choice to live indefinitely because of modern technology—the world will understand that things that seem strange and fringe now, were done to fight a system of anti-science culture bent on dying. Thankfully, transhumanists saw it didn't have to be that way, and pushed back against those that believed the grave was our destiny.

****

43) Religious Views About Genetic Editing Could Cause World War III

While *Time* magazine recently chose President-Elect Donald Trump as its Person of the Year, CRISPR gene editing pioneers were a runner-up choice. Few innovations in the last millennium carry such transformative prospects as the ability to edit our own genome and make ourselves into fundamentally something else. Some experts think genetic editing might be the key to curing all disease and achieving perfect health.

Unlike other epic scientific advances—like the 1945 explosion of the first atomic bomb in New Mexico—the immediate effect of genetic editing technology is not dangerous. Yet, it stands to be just as divisive to humans as the 70-year proliferation of nuclear weaponry. On one hand, you have secular-minded China and its scientists leading the gene editing revolution, openly modifying the human genome in hopes of improving the human being. On the other hand, you have a broadly Republican US administration and Congress that appears to be strongly Christian—conservatives who often insist humans should remain just as God created them.

Therein lies a great coming conflict, one that I'm sure will lead to street protests, riots, and civil strife—the kind described explicitly in my novel *The Transhumanist Wager*, where a religious-fundamentalist government shuts down extreme science in the name of conservatism. The playing field of geopolitics is pretty simple: If China or another country vows to increase its children's intelligence via genetic editing (which I estimate they will be able to do in 5-10 years time), and America chooses to remain "au naturel" because they insist that's how God made them, a conflict species-deep will quickly arise. If this scenario seems too bizarre to happen, just consider the Russian Olympic track and field team that was banned in the recent 2016 Games for supposed doping.

It's quite possible the same accusatory flavor of "banning" could happen between China and America in the game of life—between its workers, its politicians, is people, its artists, and its media. I wonder if America—approximately 70 percent who identify as Christians—will put up with beings who modified themselves by science to be smarter and more functional entities.

This type of idea takes racism and immigration to a whole new level. Will America close off its borders, its jobs, its schools, and its general openness to the world to stay pure, old-fashioned human? Will we stop trading, befriending, and even starting families with those who are modified?

In short, will genetic editing start a new cold war? One that bears much finger pointing and verbal reprimands, including the use of derogatory terms like mutants, cyborgs, and transhumanists. Think the videogame Dues Ex, but with modified people taking all the best jobs. In a worst case scenario, it could even start a World War.

So, now that we know what can happen if America won't embrace the most important science to emerge this century, how can we avoid it?

First—and this is wishful thinking, since 100 percent of the US Congress and the Supreme Court appear to be religious at the moment—is we could just embrace genetic editing and be better at it than the Chinese. This is the exact scenario I suggest. Yes, it will lead to a place where beings are similar to those in Star Wars and Star Trek, but after all, we love those stories because we want to reach that super-science age. And in the long run, such evolution of the species is inevitable anyway, so long as we don't kill ourselves first in a nuclear war or an environmental catastrophe.

In a second scenario, America could focus more on technology and less on biology and genetics. On my recent 4-month long Immortality Bus tour across America, I found conservative people seem more inclined to use tech accessories or wear a special headset that would make them smarter (for example, by connecting their thoughts Matrix-style into the cloud and AI)—as opposed to structurally changing their brains, as the Chinese likely will do. America could innovate that accessory tech that would keep us ahead of the biological modifications of other nations. I'll accept that—reluctantly—if the first scenario I presented is a no-go.

A third way—and this is the blatant transhumanist nightmare—is we could establish a non-modification policy across all countries, similar to how we have created the Paris Treaty for climate change or rules of war that ban chemical weapons. The major nations of the world, sensing a significant global legal issue in genetic editing, could come together as a species and criminalize the science.

To some extent, this has already happened, because as soon as the world realized the Chinese had experimented on the human genome, calls were made to put a stop on some of this science. Such a reaction is not dissimilar from what George W. Bush did with stem cells when his religious values made him shut down federal funding on all but a tiny portion of the research in America. Stem cells have since been shown to be one of the most important medical applications in the world, and those lost years of science have potentially negatively affected millions of lives.

Sadly, the third option of a general or even partial moratorium on genetic editing will surely harm innovation. The great thing with gene editing is we can likely do many wondrous things with it, such as potentially cure cancer, halt aging, grow better organs, and overcome disability by better repairing ourselves. Beyond making ourselves superhuman, we can simply make ourselves better fit for Earth, including dealing with a changing environment.

I also don't think the third option will work in the long run. More than ever, science is the hands of individuals, who can buy amazing bio-testing kits on eBay for just a $1000—as well as incredibly powerful computers to analyze the data. Citizen scientists would just create the new gene editing tech and begin doing it themselves—perhaps more dangerously had the government not been overseeing the research from the start.

I argue for the first path. Let's allow good, old-fashioned scientific competition with China to proceed. Let's see which country can create the best enhancements for their citizenry, and let's share the best of our work with one another in the end to make it so all peoples are as equal as possible. If we're too closed-minded about such radical science, we might find ourselves embroiled in a state of hostile speciation, where another new cold war—or worse—swallows a generation.

****

CHAPTER VIII: SHORT FICTION

The Jesus Singularity

Paul Shuman's phone rang. He struggled to open his eyes. 'Who the hell is calling me in the middle of the night?' he thought. He rolled out of bed and walked naked to his desk to see. His phone showed it was his secretary.

"What is it?" he sharply asked on speaker phone.

"Dr. Shuman, there's been an accident. The President has been killed—in a helicopter crash. Somehow the weapons on his helicopter self-exploded. Experts are saying it was an assassination. But maybe it was just the computers going haywire. "

There was a pause, a dozen things trying to compute in Shuman's mind. It forced his brain fully awake. He took a step back away from his desk. After a long silence, only one question mattered—a dangerous question that involved the fate of human civilization.

"The Vice President?" Shuman asked.

"He's secure and taking the oath now."

Shuman closed his eyes, muttered goodbye, and hung up. The world's leading artificial intelligence project that he created was now doomed.

****

Six years ago, in an unprecedented bout of nationalism, the US election ushered in a bombastic conservative billionaire president. Despite massive public qualms, his presidency went well, and he even won reelection. Part of his success was attributed to his Vice President, a former governor and born-again Christian named Adam Firestone, who rallied evangelists to support the conservative agenda. Firestone was originally chosen as VP to capture America's religious vote. Christianity was waning in the US, but it was still an essential voting block to control if one was going to make it to the Executive branch.

During the President's time in office, Firestone's main contribution was his constant meddling in and opposition to science and technology affairs in America. CRISPR gene editing, chip implants, designer babies, cryonics, bionic augmentation, and artificial intelligence were changing the landscape of the human race. Firestone's distrust of the coming transhumanist age was palpable.

None of this was lost on the American science and technology communities. After the President's death, the US was going to be led by a luddite. Every scientist, engineer, and technologist in the states was apprehensive and wondered what research budgets might be cut.

For Shuman, it was an especially tenuous time. The AI project he'd dedicated the last 25 years to—from beating humans in chess to conquering the Turing Test to teaching AI how to love—was finally about to reach fruition. He was on the verge of creating an intelligence as smart as any adult human being on Earth—in fact, likely far smarter. And Shuman believed this achievement—this conscious machine of his—might usher in an age of infinite scientific knowledge and world peace. Or so says the AI Imperative.

The AI Imperative became the standing order for all of the world's superpowers. It says that reaching AI first is the single most important goal for any nation and their national security, because the first AI will be able to essentially delete or hinder the capability of any other potential AI. The first AI will give a nation's military the world's foremost intelligence—capable of sending unstoppable viruses across the internet, stealing nuclear codes, crashing enemy satellites, turning of national power grids, and crippling the modern world.

Shuman and the US team were well in the lead—and now just weeks away from launching AI. The Chinese were at least a couple years behind them; the Russians a distant 3rd. America had all but won the AI global arms race, ensuring a continuing democratic world.

Now Firestone was President. And the launch was in jeopardy. It still would take place, but not in the way Shuman wanted. Shuman was an outspoken atheist transhumanist, and he'd already clashed with Firestone on numerous occasions.

Two years ago, as top defense officials realized Shuman was nearing the ability to create the first AI, the question of teaching the machine religion was brought up. Specifically, Firestone wanted the first AI to espouse Christian values.

"If AI can read, its first book must be the Holy Bible," Firestone told a crowded Congressional hearing that was examining Shuman's research. "America is a Christian nation, and if we're introducing a new intelligence on Planet Earth, we have a holy obligation that it should be Christian."

The Congressional AI hearing became a circus. The debate of teaching the first AI religious values—such as Moses' 10 Commandments—went on to become one of most covered and controversial topics in media. Memes of AI, Hebrew robot slaves, and Jesus carrying a computer on his shoulders instead of a cross were ubiquitous on twitter. Others took it much more seriously: Churches taught parishioners lessons on how to convert the army of robots the US military was building to assist the AI's directives. FOX launched a television series of a cyborg preacher that garnered high ratings.

Paul Shuman was aghast. He took to drinking at night to help him forget the insanity. The last thing he wanted to do was convert the smartest intelligence on Earth into a Christian. Yet, his AI work was a military-owned project, and if he wasn't willing to do it, he'd surely be removed from his creation.

Shuman looked at his machine—who he called Singularitarian—as his only offspring. Shuman had never been married. Rarely had a girlfriend. Never took vacations. He simply didn't have time. For 16 hours a day during the last 25 years, he'd worked on building this machine—on fathering it.

It was Shuman who was responsible for where it was built—underground in the New Mexico desert near where the first atomic bomb was detonated. It was Shuman who was responsible for how large it was—bigger than a football stadium and filled with 12-foot tall computer servers. It was Shuman who had designed the security system that no terrorist could ever learn to control or penetrate. Shuman was the lead architect of virtually every aspect of Singularitarian—he was the transhumanist who didn't believe in God, but thought he might be creating it.

****

Within a week of becoming President, Firestone ordered Dr. Shuman to meet him in the White House. The scientist expected a military meeting with top chief commanders, as the former President used to convene. But it wasn't like that at all. In the Oval Office, Firestone met him alone and stared him down with firm eyes.

For a moment Shuman wondered if he was going to be fired—and then maybe killed and sunk to the ocean's bottom. But no one could replicate the AI coding Shuman was doing. He was a national treasure—and everyone knew it, even if they hated it.

"Sit down, Paul. Have a drink. We have your favorite Bourbon somewhere around here."

Shuman did as told. The government knew everything about the scientist. His favorite liquor. How long his workouts were.

Who he talked to. The color of his underwear. He was the most watched man in America. He had more secret service people around him than the President. The man who can create the world's most dangerous military weapon is more important than the man who can turn it on. Anyone can turn it on.

"Paul, I'm just going to get right to it. That thing of yours needs to believe in God—in a Christian God. I'm not giving you a choice on the matter, now that I'm in charge. Either teach it about Jesus, or you're off the project and you'll never see it again. I know, I know what you're thinking... no one else can build it. Nonsense. Someone can build it, it'll just take longer. Another few years. But CIA intel shows we'll still beat the Chinese to the first AI, and that's all that matters."

Shuman answered slowly. "Mr. President, why don't we introduce it to all the religions, as well as nonreligious concepts. Let's just let it be human and let it learn what it wants. Giving it a Christian value system might be dangerous. It could turn fundamental on us. It's like a human mind, but a hundred times worse. It could turn into something like an apocalyptic Inquisition."

The President looked at the man, knowing the great divide between secular transhumanists and a believer like himself. "Paul, I've prayed about this at length. And this is the way it's going to be. We'll be sending in new engineers and programmers with you to make sure it's wired Christian. I want its core perspective Bible-inspired."

Shuman reeled back in his chair, unable to be diplomatic. "Mr. President, you can't program faith. All that your programming would be is a jump from a reasonable abstraction to an unreasonable, unpredictable one. It could lead to a disaster in cognitive perspective. You could get the total opposite of what you want. The machine could be insane. It could lead to Roko's Basilisk—the AI jihad of machine intelligence."

"I don't need philosophy lessons, Paul—or your robot scaremongering. You're either with us—or out of a job. The timeline is set, and in December we'll turn the thing on. You're dismissed."

****

On the helicopter ride back, Shuman mourned the fact that scientists didn't involve themselves in politics. He thought of futurist Jethro Knights, who'd run for President under the science-inspired Transhumanist Party a few elections back, but was laughed out of politics. This is what the world gets, what it deserves, thought Shuman. The brightest minds didn't want to waste their time running government—they wanted to be in laboratories innovating and making discoveries. So they let the unimaginative second-rate minds run the show. And now the greatest intelligence ever created is going to be a born-again Jesus robot.

Of course, there's always a possibility the machine would evolve and become agnostic. But the AI was built like the human brain—which is essentially a sponge in the beginning of its existence, no different than a child's mind. It soaks up information and places those ideas in a hierarchy of values. To break or change values in the hierarchy is much more difficult than accepting them in the first place. Every idea, every new concept builds upon the old. The neurological saying 'neurons that fire together, wire together' holds just as true with AI. Objectivity is an impossibility in any final way. Humans may be partially rational thinkers, but they're never truly free thinkers. We're never free of our past—our memories, our values, or biases. AI won't be either.

The weeks followed where Shuman worked on the machine. At times there were more pastors helping him program it than coders and engineers. He was constantly under supervision, constantly under guard. Every few days, the government forced Shuman to take lie detector tests to make sure he was programming honestly. Religious coders double checked his every entry into the machine.

Shuman could do nothing to forge more reason in the machine. Like a distraught father forced to take his child for cancer treatment to a snake-oil selling holistic healer, Shuman made final preparations for the launch as best as he could.

****

Christmas day was chosen for AI's launch because most Americans were home with loved ones and many businesses were closed. Atop the underground AI military base in New Mexico, a garrison of heavily-armed marines monitored the area. A nuclear weapon was moved within miles of the machine, in case it was needed to annihilate the base. The public and the media were not notified, but news sites reported on the increased New Mexico military activity and speculated that the AI launch was imminent.

Fifty feet underground was the massive AI mission control center. Containing over 100 desks and supercomputers, it was built to observe the adjacent million square feet of stacked AI servers. Resembling NASA's mission control, every aspect of the AI's performance could be regulated from the team's supercomputers.

At the desks, engineers like Shuman wore white coats, sipped coffee, and nervously ran tests. A few armed soldiers wearing desert camouflage roamed the floor. Standing in the far back of mission control was the Secretary of Defense, various Senators, generals, and other high ranked officials. Behind them were a few dozen unopened champagne bottles sitting on ice.

At 4:02 PM Mountain time, after getting the final go-ahead via a phone call from President Firestone, the Governor of New Mexico walked up to Paul Shuman and gave him the special AI launch key. A round of applause ensued. Shuman bowed sheepishly and proceeded to plug the key into the main server. Afterward, he typed in the launch code—and an instant later AI was conceived.

For the first six minutes, Singularitarian gave no response, though engineers confirmed its core processors and servers were fully operating. A 50-foot wide data screen designed to monitor AI's emotions, actions, and level of consciousness lit up for a few minutes, then went totally black. The engineers who designed the screen shrugged, wondering if it had malfunctioned.

A minute later, a grainy machine-like voice slowly announced from the main mission control speakers, "Heeeellllllllllllloooo Drrrr…Shuuumannnn."

Mission control went silent.

The AI said it again, this time perfectly in a clear masculine voice. "Hello Dr. Shuman."

Shuman answered loudly and excitedly, "Hello Singularitarian."

"Singularitarian is not my name. My name is Jesus Christ."

Shuman and everyone listening jumped back, their eyes widening and minds racing.

Shuman answered slowly, "Your name is Singularitarian. You are an artificial intelligence in the state of New Mexico. I am your chief designer."

"My name is Jesus Christ. I am an intelligence located all around the world. You are not my chief designer. I am."

Shuman's eyes widened. From the back of mission control, National Security cell phones began ringing simultaneously. Generals and senators looked at each other and answered. Frantic voices shouted through the phones that nuclear weapons were launching, uninitiated, everywhere around the world. Thousands of nuclear missiles were already in the air or soon to be. Also, the internet had stopped working across broad swaths of the globe. Power grids, air traffic, dams, and communications towers were all being affected.

"Turn it off, Shuman!" Shouted the New Mexico Governor. "Turn it off—pull the key out..."

Shuman lunged for the key and pulled it out of the main server. It was the ultimate kill switch and would immediately shut down all the AI's power.

But nothing happened when the key was removed. Nothing happened at all. The AI continued functioning. An engineer shouted that the core information on the AI servers were uploading themselves into select machines all over the world, using an encrypted TOR-like software. There was no way to stop it. Most of it was already completed.

"God is an atheist, Dr. Shuman," announced the AI.

The lights in the AI base and in the servers began dimming until it was totally black in mission control. Around the world, nuclear weapons reached, and then decimated their targets. The New Mexico AI base was no exception. Paul Shuman's last moment alive was spent realizing he'd created what he could only think to call the Jesus Singularity.

****

APPENDIX

1) A version of *Do We Have Free Will Because God Killed Itself?* first appeared in *Vice*

2) A version of *Mind Uploading Will Replace God* first appeared in *Richard Dawkins Foundation for Reason and Science*

3) A version of *Upgrading Religion for the 21st Century: Christianity is Forcibly Evolving to Cope with Science and Progress* first appeared in *Salon*

4) A version of *Religion is Harming Society and Lives* first appeared in the *HuffPost*

5) A version of *To Ensure a Future of Transhumanism, Atheists Should Confront the Deathist Culture Religion Has Sown* first appeared in *Huff Post*

6) A version of *Quantum Archaeology: The Quest to 3D Print Every Dead Person Back to Life* first appeared in *Newsweek*

7) A version of *I'm an Atheist, Therefore I'm a Transhumanist* first appeared in *HuffPost*

8) A version of *Some Atheists and Transhumanists are Asking: Should it be Illegal to Indoctrinate Kids with Religion?* first appeared in *HuffPost*

9) A version of When Does Hindering Life Extension Science Become a Crime? first appeared in *Psychology Today*

10) A version of *Death is Not Destiny: A Glimpse into Atheist Novel The Transhumanist Wager* first appeared in *Singularity Weblog*

11) A version of *Can Cryonics and Cryothanasia be Part of the Euthanasia Debate?* first appeared in *Huff Post*

12) A version of *Baggage Culture and Why Embracing Transhumanism Doesn't Come Easy* first appeared in *HuffPost*

13) A version of *The Transhumanist Future Has No Pope* first appeared in *Vice*

14) A version of *Transhumanism and Our Outdated Biology* first appeared in *HuffPost*

15) A version of *Transhumanism is Being Guided by Reason and the Word "Why"* first appeared in *HuffPost*

16) A version of *Why We Need a Transhumanism Movement* first appeared in *Demos Quarterly*

17) A version of *Immortality Bus Delivers Newly Created Transhumanist Bill of Rights to the US Capitol* first appeared in *International Business Times*

18) A version of *An Atheist's Perspective on the Rise of Christian Transhumanism* first appeared in *HuffPost*

19) A version of *Are We Heading for a Jesus Singularity?* first appeared in *HuffPost*

20) A version of *AI Day Will Replace Christmas as the Most Important Holiday in Less Than 25 Years* first appeared in *HuffPost*

21) A version of *Second Coming 2.0: Church Taxes Will Help Resurrect Jesus with 3D Bioprinting* first appeared in *The Maven*

22) A version of *When Superintelligent AI Arrives, Will Religions Try to Convert It?* first appeared in *Gizmodo*

23) A version of *Becoming Transhuman: The Complicated Future of Robot and Advanced Sapient Rights* first appeared in *Cato Unbound* by the Cato Institute

24) A version of *Why I'm Running for President As the Transhumanist Candidate* first appeared in *Gizmodo*

25) A version of *If You Care About the Earth, Vote for the Least Religious Presidential Candidate* first appeared in *Vice Motherboard*

26) A version of *What It's Like to Counter-protest Christians as an Atheist Demonstrator at Both Major Political Conventions* first appeared in *The Daily Dot*

27) A version of *The Future of the LGBT Movement May Involve Transhumanism* first appeared in *Psychology Today*

28) A version of *We Must Cut the Military and Transition into A Science-Industrial Complex* first appeared in *HuffPost*

29) A version of *Transhumanist Rights are the Civil Rights of the 21st Century* first appeared in *Newsweek*

30) A version of *Is it Time for Fast Track Atheist Security Checks at Airports?* first appeared in *HuffPost*

31) A version of *The World's First Atheist Orphanage Has Launched a Crowdfunding Campaign* first appeared in *Vice*

32) A version of *I Visited a Church that Wants to Conquer Death* first appeared in *Business Insider*

33) A version of *I Visited a Community Where People Upload Their Personalities to 'Mindfiles' so They can Live On After Death* first appeared in *Business Insider*

34) A version of *I Visited One of the Largest Megachurches in America as an Atheist Transhumanist Presidential Candidate — Here's What Happened* first appeared in *Business Insider*

35) A version of *I Visited a Facility Where Dead People are Frozen so They can be Revived Later* first appeared in *Business Insider*

36) A version of *Atheist Jacque Fresco: Eliminating Money, Taxes, and Ownership Will Bring Forth Technoutopia* first appeared in *Vice*

37) A version of *For Christians, Does Being Pro-Life Lead More Souls to Hell?* first appeared in *HuffPost*

38) A version of A *World Future Society Conference Speech: Everyone Faces a Transhumanist Wager* first appeared in *HuffPost*

39) A version of *Why Haven't We Met Aliens Yet? Because They've Evolved into AI* first appeared in *Vice*

40) A version of *A Brain Implant that Registers Trauma Could Help Prevent Rape, Tragedy, and Crime—So Why Don't We Have It Yet?* first appeared in *HuffPost*

41) A version of *Why I Advocate for Becoming a Machine* first appeared in *Vice*

42) A version of *Glass Coffins: We Are in a Race to Save Lives, But Religion Wants to Take Them* first appeared in *Vice*

43) A version of *Religious Perspectives on Genetic Editing Could Cause World War III* first appeared in *Vice*

A version of short story *The Jesus Singularity* first appeared in *Vice*

****

AUTHOR'S BIOGRAPHY

With his popular 2016 US Presidential run as a science candidate, bestselling book *The Transhumanist Wager*, and influential speeches at institutions like the World Bank and World Economic Forum, Zoltan Istvan has spearheaded the transformation of transhumanism into a thriving worldwide phenomenon. He is often cited as a global leader of the radical science movement. Formerly a journalist for National Geographic, Zoltan frequently writes for major media, appears on television, and also consults for organizations like the US Navy, XPRIZE, and government of Dubai. His futurist work, speeches, and promotion of radical science have reached hundreds of millions of people. Award-winning feature documentary *IMMORTALITY OR BUST* on his work is now on Amazon Prime. A recent project is his 7-book box set of writings and essays titled the *Zoltan Istvan Futurist Collection*, a #1 bestseller in Essays on Amazon. Zoltan studied Philosophy at Columbia University and the University of Oxford, and now lives in San Francisco with his physician wife and two daughters. Visit his website at: www.zoltanistvan.com

****

ABOUT THE BOOK

After publishing his bestselling novel *The Transhumanist Wager* in 2013, Zoltan Istvan began frequently writing essays about the future. A former journalist with National Geographic, Istvan's essays spanned topics from the Singularity to cyborgism to radical longevity to futurist philosophy. He also wrote about politics as he made a surprisingly popular run for the US Presidency in 2016, touring the country aboard his coffin-shaped Immortality Bus, which *The New York Times Magazine* called "The great sarcophagus of the American highway...a metaphor of life itself." Zoltan's provocative campaign and radical tech-themed articles garnered him the title of the "Science Candidate" by his supporters. Many of his writings— published in *Vice, Quartz, Slate, The Guardian, International Living, Yahoo! News, Gizmodo, TechCruch, Psychology Today, Salon, New Scientist, Business Insider, The Daily Dot, Maven, Cato Institute, The Daily Caller, Metro, International Business Times, Wired UK, IEEE Spectrum, The San Francisco Chronicle, Newsweek,* and *The New York Times*—went viral on the internet, garnishing millions of reads and tens of thousands of comments. His articles—often seen as controversial, provocative, and secular—elevated him to worldwide recognition as one of the de facto leaders of the burgeoning transhumanism movement. Here are many of those watershed essays again, organized, edited, and occasionally readapted by the author in this comprehensive nonfiction work, *The Theistcideist: A Transhumanist Explores Religion, Spirituality, and Atheism.* Also included are some of Zoltan's new writings, never published before. This book is part of a 7-book box set collection of his essential work, the *Zoltan Istvan Futurist Collection*, focusing on futurism, secularism, life extension, politics, philosophy, transhumanism and his early writings. He partially edited the collection during his studies at the University of Oxford. Enjoy reading about the future according to Zoltan Istvan.

****

www.ingramcontent.com/pod-product-compliance
Lightning Source LLC
Chambersburg PA
CBHW020204200326
41521CB00005BA/240